D0001633

GROUND RULES
FOR GARDENERS

A Practical Guide to Garden Ecology

Stefan Buczacki

COLLINS
8 Grafton Street, London W1

For Beverley, Julian and Simon

All the photographs in the book were taken by the author

Artists:
Ann Savage
Jayne Ivimey
Gill Tomblin
Norman Arlott

Designed by Bridget Morley

First published 1986
Copyright © Stefan Buczacki 1986

William Collins Sons and Co. Ltd
London · Glasgow · Sydney
Auckland · Toronto · Johannesburg

ISBN 0 00 219322 1

Set in Bembo by Ace Filmsetting Ltd, Frome, Somerset
Colour origination by Alpha Reprographics Ltd, Harefield, Middlesex
Printed and bound by William Collins Sons and Co. Ltd, Glasgow

ACKNOWLEDGEMENTS

This book would never have appeared without the faith in my ideas, and
the wit to understand my explanations of them, displayed first by Robert MacDonald
and then by Crispin Fisher of Collins. I owe them a very great debt. The book would
not, however, have appeared in its present form without the singular skill of my
designer Bridget Morley, who has brought her consummate talents to bear
on translating my crude ideas into a quite superb design. Together with her artists
I believe she has created a work with a very special aesthetic quality.

Finally, I must record my indebtedness to Joyce Lambert,
an outstanding British ecologist who, twenty years ago, first showed
me 'what grows where, and why'.

STEFAN BUCZACKI

CONTENTS

Chapter One

Ecology and the Garden

It is an unusual gardening book that includes blue whales, elephants and 'green' politics among its pages; that spans Californian redwoods, the science of the greenhouse, the spacing of parsnips, Capability Brown and the last ice age. Yet an encounter with all of these falls within the purpose of the gardening journey upon which you have now embarked.

Ecology, conservation and environment make up a trio of the most widely used, misunderstood and misapplied words of our age. Rarely, however, is any of them used in relation to gardens and gardening. To some degree this book is concerned with all three, and it is important therefore that I define my understanding of them. Environment is perhaps the easiest; essentially it means surroundings and, for present purposes, this is the immediate surroundings of man in the shape of his garden, although I shall on regular occasions pass beyond the garden gate to place the garden and its inhabitants in the context of the wider environment of the surrounding countryside, the British Isles and, indeed, the world at large. Conservation is 'the act of keeping entire' or 'preserving', but it is popularly used in the sense of 'leaving alone' or of 'caring for'; only the latter can have relevance to gardening however, for a garden left alone will, as I shall be discussing, soon cease to be a garden and become a wilderness. Nor can a garden be 'kept entire' or 'preserved' for, as I hope will also become apparent, it is a dynamic ever-changing place and to attempt to still its flow will be to court disaster. It may be possible to preserve individual components of the garden, but in its entirety it is never static.

It is, therefore, to explain this ever-moving and evolving nature of the garden that I turn to the third of my words, to my central theme, perhaps the most misunderstood of all. If ever a word can be said to have been discovered and done to death by the populace at large, it is 'ecology'. It now appears regularly in such newspaper headlines as 'Ecological Disaster', a catch-all phrase to describe happenings ranging from the spillage of oil by tankers at sea to planning applications for new motorways. It has even permeated politics, for while we do not (yet) have a Minister of Ecology, several countries can boast an Ecological Party, although I note that some now prefer to call themselves Green Parties; implying perhaps, that they care about plants but not animals nor soil. Oil tankers, motorways and politics are not the theme of this book; and I return for *my* meaning of ecology to the definition given to me as a student: 'What lives where, and why'. The word has, in οἶκος, a house, part of the same Greek root as economy, and implies a study of the way that household affairs are managed, be they human and financial or plant and animal. Ecology has indeed sometimes been called animal or plant sociology. Stretch your understanding of house to include the place where any organism lives and you will see the validity of my definition.

Though the evidence for ecological principles is all around us, the processes are largely unappreciated features of much that we see and comment upon. Consider first the distribution of species. Lions occur in Africa but no longer in Britain. Snakes, water voles and moles are absent from Ireland and the Camberwell Beauty butterfly has long gone from Camberwell. The lady's slipper orchid is a rare plant of northern woods. *Rhododendron ponticum* occurs very widely throughout Europe but is not native to Britain (although it grows here with vigour when planted deliberately). *Gentiana verna* does not flower as well in some places as in others. Unfortunate is the man who shrugs his shoulders and dismisses such observations as inexplicable or, worse, irrelevant. There are explanations for them all, but do the answers lie with the weather or in the soil? Are they the results of human action? Have they affected our organisms directly, or indirectly? If indirectly, what has been the medium of their operation, and do such effects explain why some organisms always occur together whereas others never do? Why does the European wild cat tend to be absent from areas where the lynx occurs, and why are red and grey squirrels seldom to be seen together? Why does the toad-stool *Leccinum scabrum* always occur under birch trees, and why do bluebells grow so profusely with oaks? If the science of ecology doesn't have the answers to all such questions, it does know how to set about discovering them, how to analyse distribution and behaviour, how to observe and record action and interaction. It recognises the subdivisions of autecology, the ecology of a single species, and synecology, the ecology of species' interactions.

Human beings tend to judge the behaviour of plants and animals by reference to the activities of one species alone, their own. Yet *Homo sapiens* is a highly unusual animal in many ways, not least in the criteria it seems to adopt for association, both within its own species and with others. Plants and animals congregate together naturally only for reasons of defence, nutrition and/or reproduction; conversely, but not inevitably, refraining from congregation may also be related to one of these underlying motives. The toadstool with the birch tree, the bluebell with

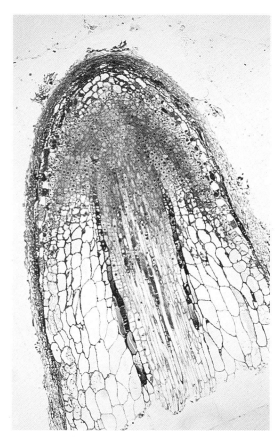

A longitudinal section, seen under a microscope, through the tip of a birch root showing the enveloping growth of the fungus Leccinum scabrum *which occurs nowhere else.*

ECOLOGY AND THE GARDEN

The cliff-top plants of the Pembrokeshire coast in spring. Thrift, squills, grasses and many other species grow in close association, although all may be found growing quite separately in other places.

Perhaps the supreme example of living things that grow in permanent and obligate association is a lichen. A common species of heaths and moors is this Cladonia *which comprises a species of fungus and a species of alga that are so intimately associated as to form a dual organism. In this, lichens are unique among life on earth.*

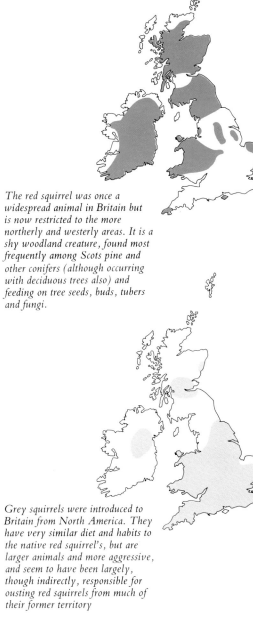

The red squirrel was once a widespread animal in Britain but is now restricted to the more northerly and westerly areas. It is a shy woodland creature, found most frequently among Scots pine and other conifers (although occurring with deciduous trees also) and feeding on tree seeds, buds, tubers and fungi.

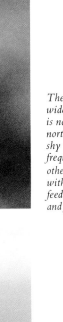

Grey squirrels were introduced to Britain from North America. They have very similar diet and habits to the native red squirrel's, but are larger animals and more aggressive, and seem to have been largely, though indirectly, responsible for ousting red squirrels from much of their former territory

Although it is usually possible to determine the reasons why animals and plants associate together, this can be immeasurably more difficult with human beings. The ecology of this American college band would tax even the most experienced biologist.

the oak wood, are respectively instances of nutritional demands and, in part, of defence. The geographical separation of wild cat from lynx, and red from grey squirrels, may also have a nutritional motive in that the different species have similar diets. But who can analyse the ecology and define the biological purpose of the football crowd, the bird watcher or, indeed, the modern suburban gardener prowling around his bed of roses? The first lesson to be learned in analysing the behaviour of other living things is not to attempt to judge them by our standards.

Ecology has always seemed to me to be a classically analytical science and I shall return time and again to the notion that to observe meaningfully is the first step to analysis, and that analysis leads to understanding. If ecology itself is a pure science, however, it has spawned applied branches that capitalise upon the understanding; and it is as well that man knows what he is doing when he attempts to manage or govern the lives of other organisms, and imposes constraints upon them, upon their environments and, by chain reaction, upon many other organisms too.

Most conventional ecological texts concern themselves with almost every habitat in which plants and animals live *except* those most influenced by man and, above all, except the garden. Generally, this absence is explained and justified by reference to the garden as an unnatural habitat, implying that it is one shaped entirely by human endeavour (and even, therefore, unworthy of study). Even if this were true, it would be wrong to imagine that the principles governing the way that organisms live no longer apply. But, more importantly, and as anyone who has ever gardened will verify, the man who thinks that he is the sole master of this particular habitat is a fool. The gardener treads a narrow line between controller and controlled; there is a limit to how far he can dictate his choice of plants, where he may grow them, when he can sow and plant them and which he may place together. For the rest, he is dependent upon the wind and sun, the soil, the rain, the hills, valleys, rivers and sea, which determine the nature of the garden itself and of the countless other organisms that populated his land before he appeared. The garden is one of the most complex of habitats and can, on occasion, almost defy analysis; and yet it is one to which many of the basic principles of ecology have seldom been applied. It is precisely to attempt this analysis that I write, for by understanding the multitude of forces that shape your garden, and enhance or frustrate your gardening activities, you will garden more effectively and most assuredly more enjoyably.

Fishermen and farmers, among others, share with gardeners the manipulation of natural habitats and their organisms. The concern of the farmer is a limited range of living things (for even a complex arable farm will contain only a handful of different cultivated plant species), and he may completely usurp the environment to achieve his ends in growing them. The fisherman, whose interest may be in one species only, can create catastrophic chain reactions by removing vast numbers of individuals. The gardener, with hundreds, if not thousands of species to consider, often seeks to grow them in harmony with the environment as a whole – and wonders why he so often fails.

The gardener is not alone in his attempts to manage other organisms; it is a role he shares to a greater or lesser extent with the farmer, the commercial grower, forester, game warden, fish farmer and fisherman. He has, however, chosen perhaps the most exacting of environments with which to work, in that the number of organisms that are his concern is huge. A typical small home garden probably contains at least one hundred different species or varieties of plant; a 'grand' garden, many thousands. And yet even the most complex of farms and forests will have only a small fraction of this diversity. Moreover, because the home garden is relatively very small in area, all its components and all its features will be magnified in their effects; a single tree will be the dominant feature of most gardens, but it will be lost totally on the farm or in the forest. Even more significantly, any shortfall in weed control will be very much more apparent and serious in a home garden than ever it will among rolling acres. I concede that the gardener does not have his livelihood at stake in the appearance and performance of his garden; but measured in terms of the importance to him of each unit area of land, his need for a grasp of ecological principles is as burning as anyone's.

Although inadequate weed control measures sometimes reveal themselves conspicuously (and very attractively) on farmland, shortcomings in reducing the competition that cultivated plants must face are generally much more evident within the confined area of a garden.

Left *Commercial forestry utilises monocultures or, at best, a very small number of species and grows them on what, to a gardener, must seem a very long rotation indeed – about thirty years before felling is the average age attained by most softwood trees cultivated in Britain.*
Right *The relative stability or relatively uniform rate of change of an existing ecosystem can be upset dramatically by the introduction or natural spread of a highly competitive species – in this instance, the rosebay willowherb* Epilobium angustifolium *has swamped much of the original vegetation by its ability to colonise scrub land rapidly.*

In succeeding chapters, I shall discuss in turn the various ecological influences of which a gardener should be aware, but one important general principle is best introduced here. I have referred to the chain reaction that occurs whenever we impose our will upon a natural habitat – when we disturb what has popularly become known as the balance of nature. To illustrate the complexity of the disturbance, I find the analogy with a balance, in the literal sense, to be appropriate, for the very expression 'balance of nature' conveys the impression of a fairly stable situation. Touch a balance pan with a finger and it swings to and fro before eventually coming to rest again. Place a weight on one pan and find in how gradual and piecemeal a way you must counter this by small additions to the other pan, if the whole is not to swing violently or fall from its pivot. Now imagine a multi-directional balance, a Heath Robinson construction, where one small addition to one small pan has far reaching and consequential disturbances throughout the entire system and you gain some impression of how delicate are the relationships between the many organisms in your garden. Place in the midst of a more or less stable environment any new organism and the existing population of organisms and the environment itself (known collectively as the ecosystem) will react to it. The scale of the reaction will depend on many factors. Often, organisms physically small and few in number will have less dramatic effects; the ecosystem may be buffered sufficiently to absorb them. Release one wood mouse into a garden and the consequences will be small. Introduce an elephant and they could be catastrophic. Or, at least, catastrophic in the short term: provided the elephant doesn't take up residence in the garden, the plant life will regenerate in large measure within a few years. It will be a rather different environment, but it will recover. Nonetheless, size and number are not always the best criteria to use in predicting size and duration of impact. The myxomatosis virus is very small and was introduced to Britain in very small numbers. It most effectively removed rabbits from large areas of grassland, however, and completely changed the character of parts of the British landscape in ways that I shall examine in due course. Such consequential effects have been called chain reactions and the most familiar examples are those that centre upon nutrition, food chains. Even the most elementary of school textbooks has such examples as man eating cow that, in turn, has eaten grass. In an ecosystem, life and food chains are very much more complex. They are also, like my Heath Robinson balance, multi-directional and are better described as webs than as chains. Diag. p. 17 gives an example of just such a food web, but it should be thought of as rather more than a representation of A eats B, and B and C eat D, for the whole purpose of nutrition in any organism is to release energy. Thus, the food web is an energy web, chronicling the extremely efficient way that an ecosystem, through its component organisms, makes maximum use of the solar energy trapped by green plants. Relationships like these between organisms and their environment occur throughout biology and so in Diag. p. 17 I have not named the individual organisms, but have instead merely shown the flow of energy, and of biological matter and essential gaseous materials, through a stylised ecosystem. This diagram could, in part, represent a South American rain forest, an Arctic tundra, a field of sheep, or your garden. For your garden is indeed an environment of interacting organisms, of which you are, or pretend to be, the most important. It is a highly complex and, in a biological sense, a noble place, worthy of your careful attention. It is where life and death continues at all levels and at all times, influenced by and responding to each and every action that you take. Thus forewarned and forearmed with this knowledge, we can begin a piece-by-piece examination of this quite fascinating habitat and learn how to manage it more effectively.

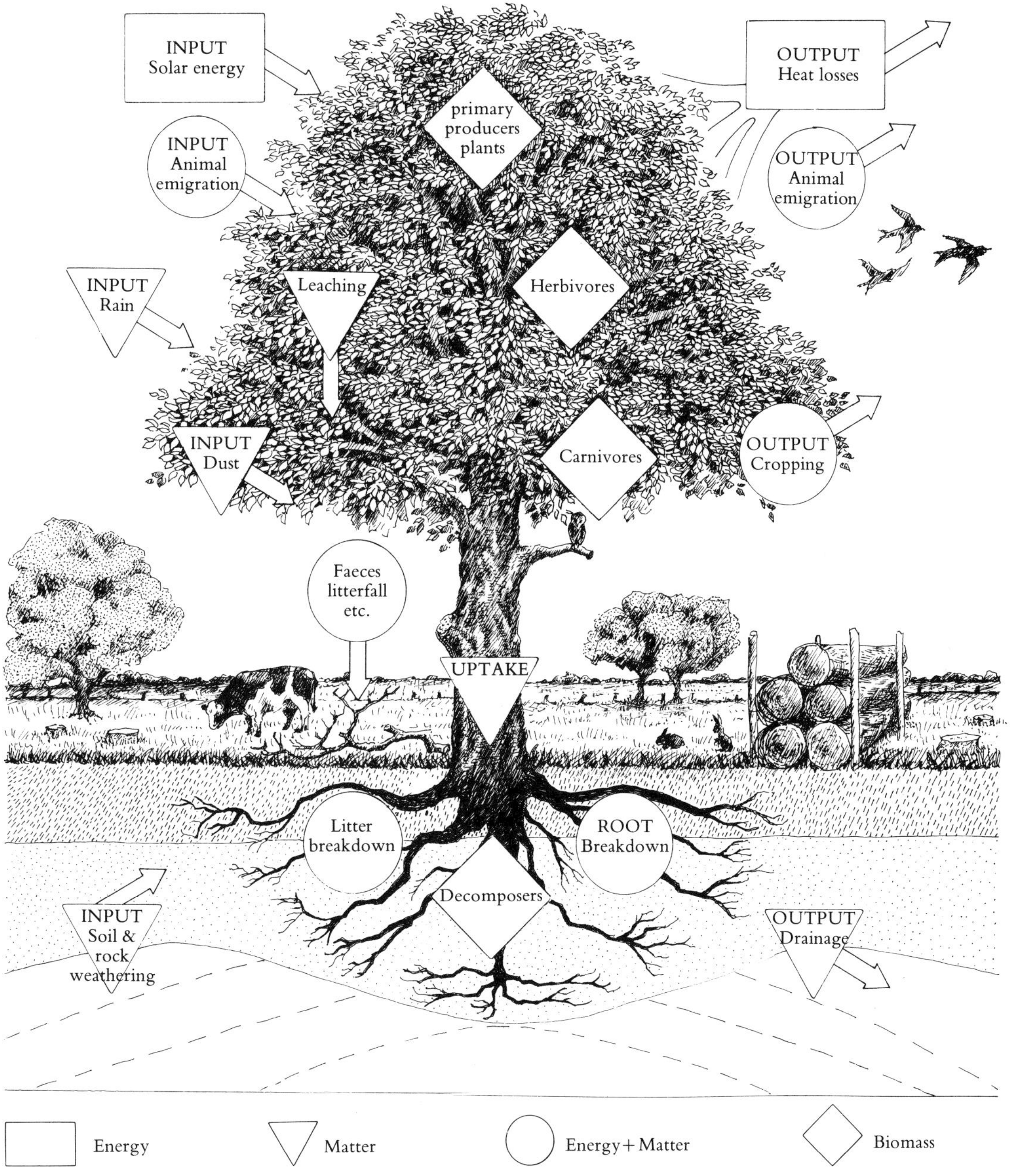

The dynamic and ever-changing blend of animate and inanimate objects that is a natural ecosystem contains countless small inter-reactions of them all. Some of the larger, more readily recognisable trends are shown here, where the flows of energy and of organic and inorganic matter into and out of the system can be identified, together with the major living things which constitute the more or less permanently present biomass that depends on, is influenced by and contributes to all of these factors. On a diagram of this scale, only the very broadest trends can be displayed; but clearly the removal of any one of the elements shown will have a marked effect on those remaining.

Although the diagram is based loosely on a modern British farming environment, comparable inputs and outputs will occur in other ecosystems, some being more prominent in certain systems than in others.

(Diagram based on one by J. P. Ovington)

INPUT Solar energy

INPUT Animal emigration

INPUT Rain

INPUT Dust

primary producers plants

Leaching

Herbivores

Carnivores

Faeces litterfall etc.

UPTAKE

OUTPUT Heat losses

OUTPUT Animal emigration

OUTPUT Cropping

Litter breakdown

ROOT Breakdown

Decomposers

INPUT Soil & rock weathering

OUTPUT Drainage

☐ Energy   ▽ Matter   ○ Energy + Matter   ◇ Biomass

Chapter Two

Climate, weather and the garden

The British weather has been blamed for most facets of our national life and is undeniably the greatest single environmental influence on British gardening. Within such a small geographical region, the great variability of climate and weather, without real extremes of temperature, precipitation and wind, probably permits British gardeners a wider choice of plants than it would be possible to grow within an area of comparable size anywhere else in the world. And yet, that same British weather, in its great unpredictability, presents a gardener with enormous challenges and frustrations. Allied to this variability and unpredictability is the lack of clear boundaries between seasons; such calendar points as 'the first day of Spring' or the fourth Sunday after Michaelmas have never been of much value as signposts for gardening activity!

The complex nature of the British climate, and the difficulties in forecasting its weather, are explained by the geographical situation of the British Isles – close in the west to a large ocean mass, and in the east to a huge continental region, while lying only around 30° south of the North Pole. Air movements from the Euro-Asian land mass, which is consistently dry, but cold in winter and warm in summer, tend regularly to depress and raise our temperature in these seasons. On the contrary, the warm water current generally known as the Gulf Stream, tends to warm up the British Isles in winter and cool it in summer, often (being a maritime influence) to the accompaniment of rain and clouds. Really unseasonable cold or warm spells of weather, on the other hand, are often due neither to the continent nor the ocean, but to the penetration of cold polar air southwards or of warm tropical air northwards.

Although British seasons do merge with one another, there is no denying that we have a winter; a season when frosts are frequent and temperatures in general are low. (Clearly this is not like being in the Tropics, where seasonal differences, if they exist at all, are more in terms

A garden can appear so different from one season to another that 'landmarks' in the form of evergreen plants, the overall shape of deciduous trees, or the non-living features of the environment are needed to facilitate our identifying these two pictures as of the same place. Responsible for, and adding to, these gross differences are countless unseen happenings, in the soil, in the air and within the plants themselves.

of rainfall than of temperature.) This means that we have a genuine resting or dormant season, when most plant growth slows down or virtually ceases altogether. Deciduous trees and shrubs shed their leaves, herbaceous perennials die down to a rootstock, bulb, corm or tuber, while annuals survive simply as seeds. Such dormant states are induced by lower temperatures and/or shorter day-lengths, acting through the physiology and biochemistry of the plant. Dormancy means that the plants do not exhaust themselves in the way that roses, for instance, often seem to do in the Mediterranean region, which lacks a real, frosty winter. Dormancy means, moreover, that little damage is likely to be caused if plants are transplanted at such times.

Water The importance of water to plants scarcely needs emphasising. Approximately 90% of the protoplasm of actively growing plant cells is water, and most of a plant is protoplasm. It is, moreover, in aqueous solution that mineral nutrients are taken up by the plant from the soil and in which they and the complex chemicals manufactured by the plant itself are moved (translocated) from one region of leaf, shoot or root to another. Deprive a plant of water and it will die from a combination of starvation, structural instability and general biological malfunction brought about by the irreparable damage caused to cells through prolonged loss of turgidity – the fully hydrated state in which they maintain a more or less constant volume and shape. The uptake of water by plant roots, its upward translocation in the type of conducting tissue termed xylem, and its loss by transpiration through the leaf pores (stomata), comprise the transpiration stream; and although a little water uptake is possible through plant leaves (a fact made use of in foliar feeding), this is of minimal importance in nature compared with uptake from the soil by roots. The amount of water present in the soil at any particular time depends on the physical characteristics of the soil, the prevailing temperature, the drying effect of the wind and, of course, the volume of precipitation that has fallen on it (see Chapter 3).

Precipitation is a useful term for all the forms in which water falls to earth from the atmosphere. Most falls as rain, although drizzle, dew and fog also contribute small amounts of liquid water; hail, snow, sleet and hoar frost add greater or lesser volumes of more or less solid water.

Certain combinations of circumstances, or of topographical and geographical features, will render rainfall likely if moisture-laden air is present. There are obvious regional variations in annual rainfall (see Map p. 22) and, in general, proximity to ocean masses and the presence of high ground mean more rain. In Britain, this is mainly the result of warm westerly or southwesterly winds passing over the Atlantic Ocean and causing water to evaporate. When such moist air reaches the high land of the west and central parts of the country, it rises, cools and expands. The moisture condenses, clouds form, and rain falls. This is relief or orobatic rain.

There are two other types of rainfall, however, of which gardeners should be aware, though much less constant and predictable in their occurrence. Cyclonic rain is produced in the low pressure systems so familiar on weather maps. Here, warm and moist air from the south, blowing northwards, rises above cold and dense polar air moving southwards. As with relief rainfall, the rising moist air expands and cools, but with the significant differences that cyclonic rain can occur over low as well as high ground and that, as low pressure systems unlike mountains are constantly moving, outbreaks of such rain are much harder to predict.

Finally, the least predictable of all is convectional or thunder rain. Here, localised heating of the earth gives rise to powerful upcurrents of warm, moist air. The result is towering clouds, up to 6,000 metres high, in which large raindrops form, buoyed up by the rising air. Lightning often accompanies such rainfall and may strike to earth and damage trees or ground vegetation; but only the presence of large and valuable trees, close to buildings, really justifies the fitting of lightning conductors. The very heavy, short-lived rainfall of thunderstorms can have considerable ecological impact in gardens through sheer physical force: flowers and vegetation are beaten down or broken, and the soil may become compacted and capped. More insidious, however, is the enhanced splashing upwards of soil onto low-hanging fruits, for this soil commonly contains micro-organisms able to cause decay.

Lack of rain, drought, is of course more important in an ecological and gardening context than an excess; in Britain, a drought is considered officially to be a period of 15 days when no measurable rain has fallen. There are regional variations (Map p. 23) but the occurrence of drought is hard to predict. Although plants differ in their ability to tolerate drought and in the periods of growth when they benefit most from additional watering, the use of hose-pipes and taps is often controlled officially during drought periods. Thus the improvement of the soil's water-retentiveness assumes great significance (p. 66).

The proportion of the annual precipitation that falls as snow varies across the country, but in the lowlands, even in areas such as north-east Scotland where snow falls on average on 35 days each year, the amount of moisture contributed in this way is rarely great. The fact is that 25 cm of snow represents only 2.5 cm of rain; and most, of course, falls at times when plants have little use for water. A blanket of snow, although itself frozen, will nonetheless protect underlying plants and soil; a covering of about 75 mm will usually prevent plants from being frozen. When falls of snow are very heavy, however, its sheer weight will break branches of shrubs and trees, particularly those evergreens that present a large surface area on which the snow can settle.

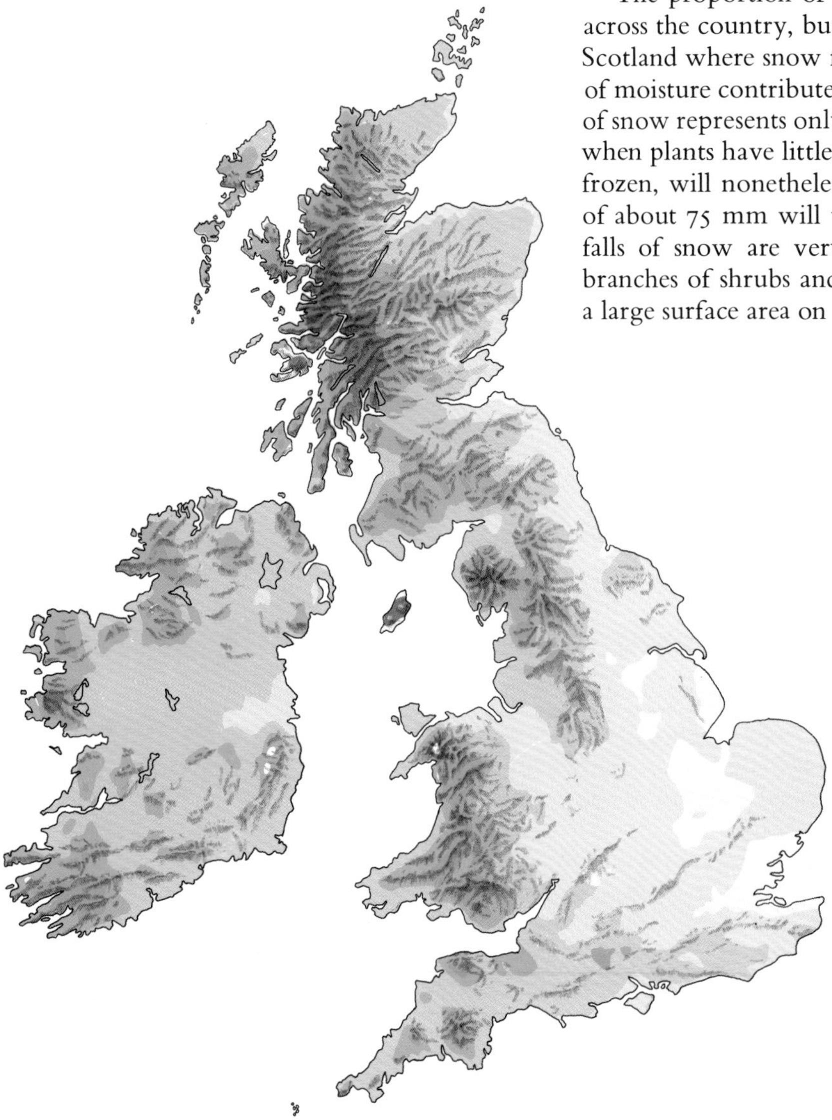

British Isles: average annual rainfall
The British Isles display the influences of high land and proximity to the ocean in the regional distribution of rainfall. While periods of heavy rain can occur even in the driest parts of East Anglia, these are of the types known as cyclonic or convectional rather than the relief rainfall which renders the north and west so consistently wet. While cyclonic and convectional rain can occur anywhere over low or high ground, the mountains that cause the moisture-laden clouds to rise and deposit their contents as relief rainfall are firmly rooted in their positions. It is because the prevailing winds have passed over such extensive areas of ocean that they are so moisture-laden.

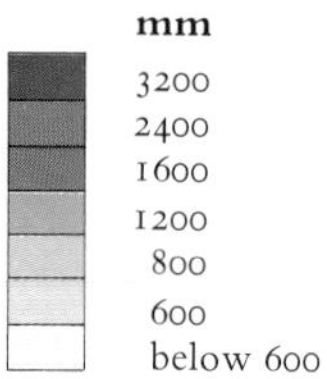

mm
3200
2400
1600
1200
800
600
below 600

Where there are very strong vertical air movements within a cloud, small accumulations of water and ice move repeatedly up and down, growing ever larger as they do so, and falling eventually as hail. Hail showers or storms occur in most areas every year and can inflict significant damage; fleshy-leaved plants such as onions and ornamental bulbs display characteristic white flecking (on the windward side only, unlike the more uniformly dispersed disease lesions), and thin-barked trees such as maples can have large areas of bark torn away.

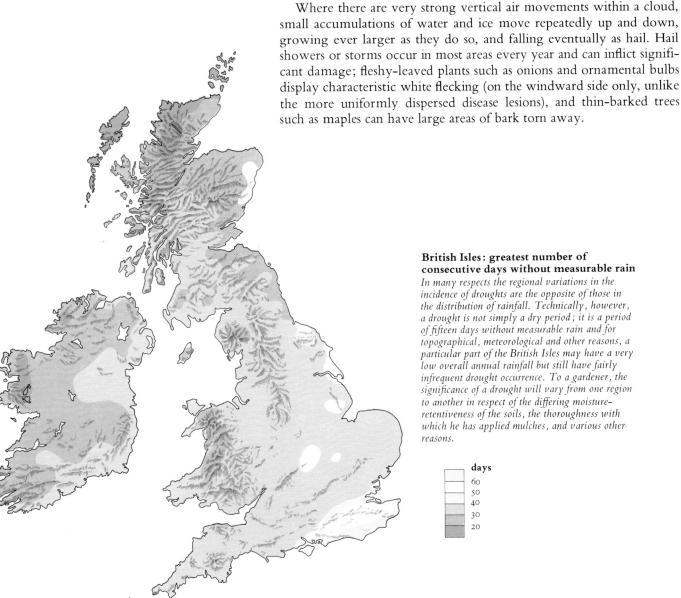

British Isles: greatest number of consecutive days without measurable rain
In many respects the regional variations in the incidence of droughts are the opposite of those in the distribution of rainfall. Technically, however, a drought is not simply a dry period; it is a period of fifteen days without measurable rain and for topographical, meteorological and other reasons, a particular part of the British Isles may have a very low overall annual rainfall but still have fairly infrequent drought occurrence. To a gardener, the significance of a drought will vary from one region to another in respect of the differing moisture-retentiveness of the soils, the thoroughness with which he has applied mulches, and various other reasons.

days
60
50
40
30
20

Temperature

The temperatures experienced in any region depend primarily on the angle at which the sun's rays strike; thus, at the Poles, where this angle is shallow, the warming effect is much less than at the Equator where the sun is higher in the sky. Even within the British Isles, there are considerable north–south differences in temperature. The enhancement of temperature by trapping the sun's warmth in cloches and greenhouses, or actually adding to it by artificial heating, is vitally important for many garden plants (Map p. 26).

All life on earth depends on solar radiation. Less than half of this radiation actually reaches the earth's surface, the remainder being lost, mainly by reflection back into space. For plants and gardens, there are two important components of the radiation: the predominantly illuminating shorter wavelength visible light, and the warming longer wavelengths tending towards the infra-red. While neither actually penetrates the soil, the surface-warming effect of the sun's heat is conducted into the top few centimetres of soil; and the re-radiation of this into the air is a very important factor in governing the temperatures actually experienced by plants. Most of the long-wave radiation leaving the earth's surface during the night-time cooling period does not actually pass back into space; it is largely retained through the absorbing properties of atmospheric water-vapour and carbon dioxide. But if the skies are clear, especially in winter when the compensation of day-time warming by the sun is low, the heat loss from the earth can be sufficient to cause temperatures at ground level to drop below freezing and form a radiation frost. A ground frost occurs when the temperature at or just above the ground has fallen to below 0°C, but this is not usually of much consequence to the parts of the plant at that level. An air frost, however, which occurs when the temperature is 0°C or less at a height of 120 cm above the ground, can be much more serious. It is not generally damaging to plants during their winter dormancy period; but a spring frost, occurring after growth has begun, can be disastrous. The average date of occurrence of the last spring frost varies regionally (Map p. 25) and is related to the length of the night cooling period. In addition, certain site features can render such frosts more likely: a dry, free-draining soil, for instance, is a poorer conductor of heat than a wet one and there is less upward transfer of heat from reserves at depth in the ground to resist a frost on the surface. Thus, such dry soils are more likely to experience radiation frosts.

A good cover of vegetation, such as grass, can limit the amount of heat lost from the soil's surface, but ironically, in keeping the soil warmer, it causes the air immediately above the grass to cool down. Hence trees and shrubs growing in lawns are more liable to spring frost damage than those in bare soil.

It should be mentioned that frosts also occur when cold air is transported over land; but since such frosts only occur in the winter, when plants are dormant, they are of much less significance to gardeners than radiation frosts.

The actual mechanisms of frost damage to plants are complex and imperfectly understood; but they contain the key to what determines

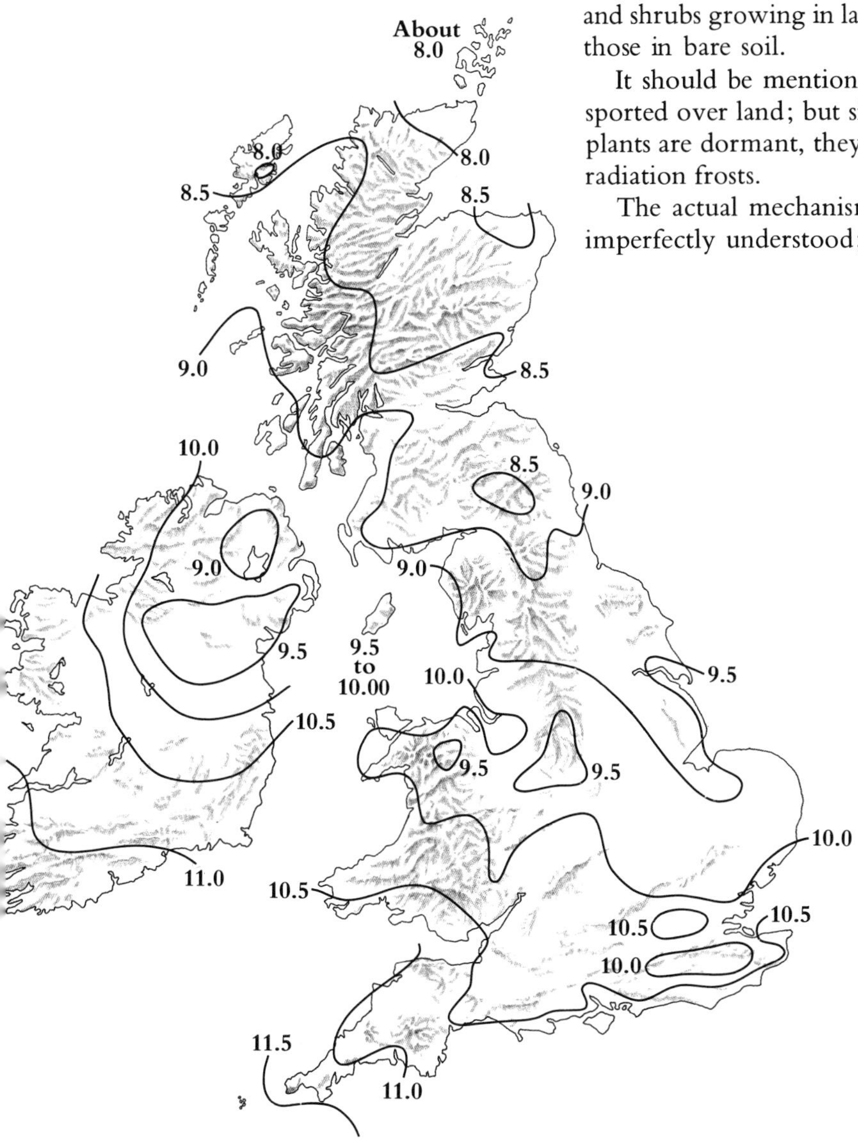

About
8.0

8.0
8.0
8.5
8.5
8.5
9.0
8.5
10.0
8.5
9.0
9.0
9.0
9.5
9.5
9.5
to
10.00
10.0
10.0
9.5
10.5
9.5
9.5
11.0
10.5
10.0
10.5
10.5
10.5
10.5
10.0
11.5
11.0

British Isles: mean daily temperature
There are many ways of expressing the range of temperatures in different parts of the British Isles – average daily or monthly maxima and minima, or daily or monthly overall averages, for instance. None is ideal, for each measure reveals a different pattern and different characteristics. That shown here is the mean daily temperature in °C. This means the average temperature of all days in the year and at least demonstrates which areas are consistently cool or consistently warm. It cannot of course take account of the differences between those areas that have a wide fluctuation about this mean and those that are much more equable.

hardiness. This is generally thought to be dependent on the ease with which water can be withdrawn from plant cells and frozen in the inter-cellular spaces, and to the rapidity with which subsequent thawing takes place. A slow thaw, allowing the cells to rehydrate gently, is much less damaging than a rapid one.

Maximum and minimum temperatures for growth are features of all plants and, clearly, they are higher for tropical than for polar species. Although all green plants grow in the same basic biochemical way, it is evident that mechanisms have evolved that permit particular species to survive satisfactorily in extreme conditions. Nonetheless, survival may

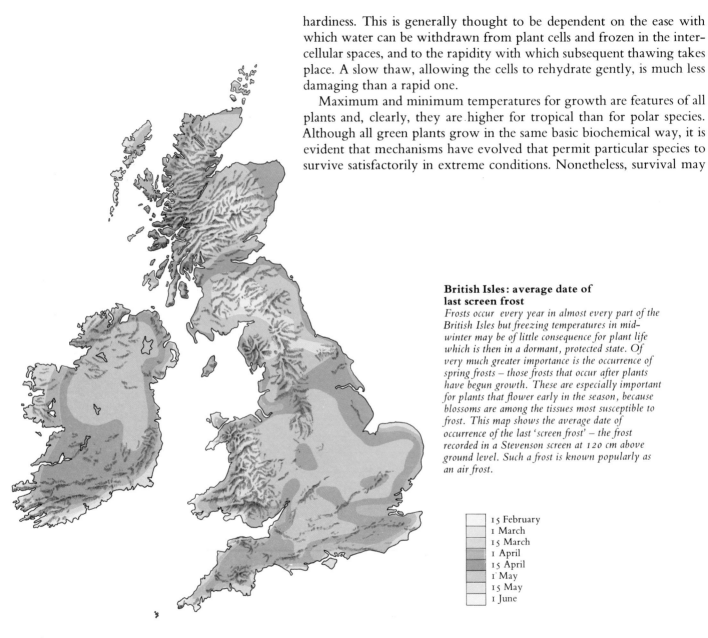

British Isles: average date of last screen frost
Frosts occur every year in almost every part of the British Isles but freezing temperatures in mid-winter may be of little consequence for plant life which is then in a dormant, protected state. Of very much greater importance is the occurrence of spring frosts – those frosts that occur after plants have begun growth. These are especially important for plants that flower early in the season, because blossoms are among the tissues most susceptible to frost. This map shows the average date of occurrence of the last 'screen frost' – the frost recorded in a Stevenson screen at 120 cm above ground level. Such a frost is known popularly as an air frost.

15 February
1 March
15 March
1 April
15 April
1 May
15 May
1 June

not be synonymous with growth and it is generally true that almost all plants grow little below 5°C or above 25°C and grow best between 10°C and 25°C. Consequently, plants in very cold or very hot climates (alpines and cacti, for instance), are usually slow growing.

Light   Light permits photosynthesis to take place and thus enables plants to manufacture food, and to develop (see Chapter 4). It is not always true, however, that the more light a plant receives, the better it will grow. Very high light intensities can be damaging, and most gardeners will be familiar with the development of greenback symptoms on tomato fruit growing too close to greenhouse glass. As with temperature and other factors, each type of plant has an optimum light requirement: plants that are naturally shade species, such as many camellias, rhododendrons or mahonias, will not thrive even in our latitudes when planted in the full sun that others prefer. That different plants have differing requirements for or tolerances of light intensity is the major factor contributing to the structure of a natural plant community – such as the ground flora, field, shrub and tree layers of a woodland (see p. 101) – and it is one that gardeners must appreciate when making mixed plantings.

A quite different role for light is seen in the phenomenon of daylength. Chemicals in a plant are receptive to light above or below certain threshold values, to which they respond by triggering physiological processes. For gardeners, this phenomenon of photoperiodism is more of interest than practical application. It is, however, of immense importance commercially where, by adjustment of the length of the period of artificial illumination given to such plants as chrysanthemums or poinsettias, it is possible to induce flowering at all times of the year. The retention of leaves by urban trees growing close to street lights is a common example of another photoperiodic effect.

Wind   The final climatic factor to consider is wind; the movement of air over the earth's surface. While wind at any one site may blow from any point of the compass, there are general, global wind patterns, and the British Isles lie in the path of the westerly air-stream of the North Atlantic. Thus, our winds blow predominantly from the south-west and, consequently, the windiest areas are the coastal regions to the north and west (see Map p. 28). The role of wind as a carrier of moisture has been covered in the account of precipitation, but dry wind is equally significant in gardens. It is the medium for the dispersal of pollen, of many seeds (of weeds, in particular, as gardeners will be aware), of insects, of spores of pathogenic fungi, of salt spray from the sea and, on occasion, of noxious artificial chemicals. The drying effect of wind on leaves results in the evaporation that has its consequences in the transpiration pull of water up plants from the soil. In extreme circumstances, the drying effect can have damaging consequences if the water cannot be replaced sufficiently quickly. Wind is, moreover, a powerful physical force and can be seriously damaging in gardens; gales may smash greenhouses, uproot trees or at least break branches, while lesser strengths can rock shrubs to and fro or remove fruit from trees. Protection from the wind is necessary to some extent in most gardens.

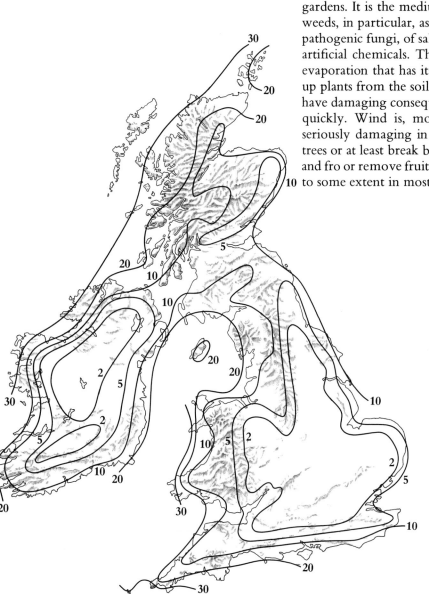

British Isles: average annual number of days with gales
There are several ways of expressing wind strengths on a map; but, as gardeners are only really interested in the incidence of strong winds, the map chosen depicts the average number of days per year on which gales have been recorded. A gale is a wind speed in excess of about 35 knots and, as may be expected, the regions most frequently experiencing such conditions lie to the west of the British Isles where winds that have blown without interruption across many hundreds of miles first 'strike land'. Nonetheless, even within very small areas, too small to be shown on the map, there will be variations in wind strength with varying exposure.

Managing the weather
The importance of site and aspect

The title of this section may seem unduly pretentious, but if a gardener can't actually influence the weather he can certainly adopt measures to minimise or enhance the effects of it on his garden. He must firstly understand, however, the way in which the situation and aspect of each garden imposes its own restrictions. Of over-riding importance is the geographical position within the country, which will reflect above all the prevailing temperature and hence the relative hardiness of the plants that will flourish or be expected to bear flowers and fruit outdoors there.

Gardens exist everywhere in the British Isles that men have made their homes. The variation, nonetheless, is vast, and a Hebridean cabbage garden will never support the types of plant that flourish beside a cottage under the South Downs. One factor, above all, determines this; all other regional climatic factors take subordinate place to the minimum annual temperature. St Mary's, in the Isles of Scilly, for instance, has, at 844 mm, roughly the same annual rainfall as Braemar, 800 km to the north and, indeed, its average monthly maximum temperature is, at 16.7°C, actually below Braemar's 17.0°. It is the average monthly *minimum* temperature of 5.6°, (in effect a frost-free climate), that means that sub-tropical plants will grow outdoors all year round in the Scilly Isles, whereas only the hardiest species will survive the average monthly minimum of − 5.2° in Braemar where only the month of July can at all reliably be considered free from frost.

The next facets to examine are the relative shelter of a garden and the direction that it faces, both of which influence the amount of sunshine that it receives. If a garden is on a slope, then the influence of direction is self-evident; but remember that in town gardens, or even moderately sized rural ones, the matter of whether the house is at the northern or

A garden liable to flooding
When this house was built, its situation may have been very different. The area either side of the river has sunk, because of land drainage, with the result that the river is elevated and contained only by earth banks. Many such examples can be found in the East Anglian fens.

The importance of aspect
Looking upwards as well as downwards should be an instinctive procedure for any gardener contemplating a new site. The enormous gardening differences between two otherwise similar gardens on opposite sides of a steep valley are due entirely to the particular times in the day when they are exposed to the full sun, and for how long.

A classic frost hollow
As cold air is denser than warm air, it will roll unseen down a slope and accumulate at the bottom. It is sometimes possible to plant trees, shrubs or hedges in such a way as to divert the flow of this cold air, or even to open up a barrier on the down-slope side to facilitate its continued movement away from the garden.

southern end can be highly significant. Instances are known where a garden on one side of a city street can produce good outdoor tomatoes, while on the other they never ripen. The aspect of a garden will also influence how much protection it gets from the wind, and gardeners should be aware of the position of buildings other than their own house and of trees (deciduous or evergreen) and the likelihood of their remaining as permanent features. The situation of a garden in relation to any marked contour variation will be important for all the above reasons, but also because of the likelihood of flooding (is there a river close by, or a water table near the surface, that will rise after heavy rainfall?), of drought (is the garden on a slope and the soil very free-draining?), or of severe frost (is the garden situated in a hollow or frost pocket in which dense cold air will collect?). Features related to garden siting other than those directly attributable to climate will be considered more fully in later chapters, but two examples are shown of sketches for garden designs where major choices have been dictated by climatic considerations.

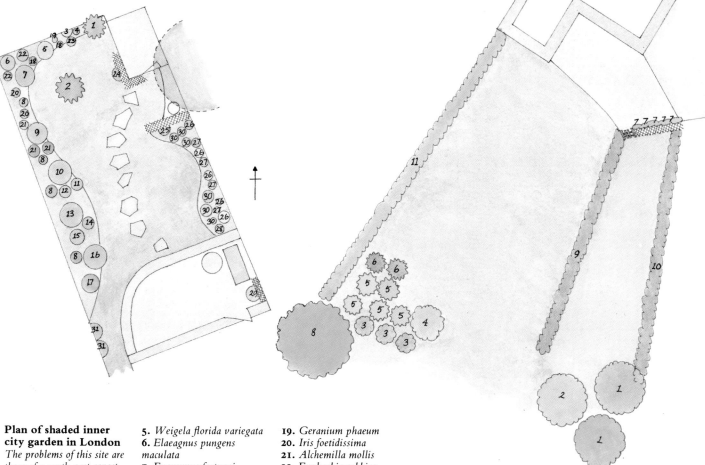

Plan of shaded inner city garden in London

The problems of this site are those of a north-east aspect, an old garden with neglected and impoverished soil, dryness at the bases of the walls and fences, existing large trees in neighbouring gardens, but relative shelter. Plants were selected to take advantage of available sunshine and for tolerance of dryness, but with sufficient colour to provide interest all year round.

1. Crab apple [existing]
2. *Juniperus scopulorum* 'Skyrocket'
3. *Hedera canariensis*
4. *Cyrtomium falcatum*
5. *Weigela florida variegata*
6. *Elaeagnus pungens maculata*
7. *Euonymus fortunei*
8. *Hydrangea petiolaris*
9. *Choisya ternata*
10. *Arundinaria viridistriata*
11. *Chaemaecyparis lawsonia* 'Ellwood's Pillar'
12. *Chaemaecyparis lawsoniana* 'Aurea Densa'
13. *Spiraea japonica* 'Goldflame'
14. *Thuja occidentalis danica*
15. *Cornus alba elegantissima*
16. *Juniperus horizontalis* 'Emerald Spreader'
17. *Acer japonicum aureum*
18. *Pachysandra terminalis*
19. *Geranium phaeum*
20. *Iris foetidissima*
21. *Alchemilla mollis*
22. *Euphorbia robbiae*
23. *Omphalodes cappadocica*
24. Clematis 'The President'
25. *Clematis montana* 'Elizabeth'
26. Rose 'Josephine Bruce'
27. *Lavandula nana alba*
28. *Geranium sanguineum*
29. Clematis 'Nelly Moser'
30. *Geranium grandiflorum plenum*
31. *Hedera helix hibernica*

Plan of exposed coastal garden in S-W England

The problems of this site are those of exposure to the south-west wind, the salt spray (the sea is only 2 km from the site), the presence of three old trees which cannot be removed because of the protection they afford, the desire of the owners to grow vegetables and a few summer flowering annuals but who have no wish or time for a great deal of gardening as they are preoccupied with a farm.

Plants were selected for wind and salt tolerance above all, therefore.

1. *Pinus nigra* [existing]
2. *Arbutus unedo*
3. *Olearia × haastii*
4. *Tamarix pentandra*
5. *Hebe* cvs.
6. *Pyracantha rodgersiana*
7. *Quercus ilex* [existing]
8. *Escallonia macrantha* hedge
9. Hawthorn hedge [existing]
10. Hawthorn hedge [existing]

Recording the weather

While it cannot be pretended that making your own weather measurements will provide you with the ability to make your own forecasts, there can be no denying that it is an interesting and instructive exercise and, over a number of years, will build up a fund of expertise that can only help the understanding of your garden and its ecology. Knowing its micro-climatic characteristics will enable you to form a picture of the ways in which it is typical of your area as a whole, by comparing your measurements of temperature, rainfall and other elements with those issued by your local meteorological office and published by your local newspaper or radio station. Thus, a general local warning of a slight air frost may enable you to judge if your garden is very likely or

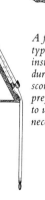

A combined maximum and minimum thermometer, known as Six's thermometer: the most convenient pattern for garden use. It comprises a U-tube containing spirit, a mercury piston and two metal pins. As the temperature rises, the spirit expands and pushes the mercury around the 'U'. One of the pins is pushed ahead of the mercury and then held at the extreme position reached. As the temperature falls and the spirit contracts, the mercury pushes the other pin to the opposite extreme where it too is held. The pins thus indicate the maximum and minimum temperatures reached until they are re-set with a small magnet.

Left *Wet and dry bulb hygrometers*: two identical thermometers, one with its bulb kept permanently wet by means of a wick and a small water reservoir. The difference between the two temperature readings enables the relative humidity to be determined by reference to tables.

Right *Earth thermometer*; designed to be set at a predetermined depth in the soil with the scale protruding above, so that it may be read in situ.

A fairly sophisticated type of sunshine recorder, an instrument which records the duration of sunshine by scorching a trace on specially prepared card. Fascinating to use, but scarcely a necessity.

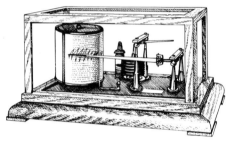

Above *Barograph*: a recording barometer which traces on a paper chart the pressure changes over a seven-day period. Thus, the passage of depressions and anticyclones is visible in a much more informative way than can be shown on the simple anaeroid barometer hanging in the hall.

Above *Rain-gauge*; a simple conical funnel on a cylindrical base which collects precipitation in an inner can. This is removed and the rainfall measured daily with the measuring cylinder.

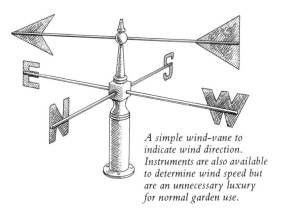

A simple wind-vane to indicate wind direction. Instruments are also available to determine wind speed but are an unnecessary luxury for normal garden use.

very unlikely to suffer, depending on whether it is relatively exposed or sheltered compared with the surrounding region. It may well be that a good gardener will learn such things by accumulated experience; but the ability to quantify matters can be even more satisfying – how much rain need fall before your soil becomes water-logged, or how many degrees of frost will a *Ceanothus* tolerate?

The most basic item of recording equipment is a minimum thermometer (or two, if you have a cold frame; or three, a greenhouse). Next is probably a maximum thermometer, and a barometer or barograph; then a rain gauge, wet and dry bulb hygrometers and, finally, a wind direction indicator, and a sunshine recorder. More useful still would be the inclusion of a small, specially constructed box (known as a Stevenson screen) to house the thermometers, and a soil thermometer. Several firms now sell basic meteorological equipment suitable for the gardener and a complete outfit (excluding the sunshine recorder), can be obtained relatively cheaply.

Weather forecasting and the gardener

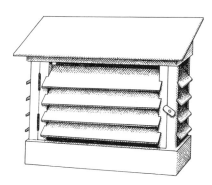

Stevenson screen; a specially designed box to house thermometers and hygrometers and shield them from precipitation and direct sunshine while allowing air to circulate freely around them. These screens are available in relatively inexpensive kit form.

Entire books have been devoted to the folk-lore of the weather and the English language is replete with such colourful imagery as 'Red sky at night, shepherds' delight' or, 'In the decay of the moon, a cloudy morning bodes a fair afternoon'. While many of these sayings do have some basis in scientific meteorology, few gardeners these days have enough faith in them to risk their plants on the strength of such guidance alone. It may be equally true that some have little more faith in the outpourings of the Meteorological Office, but few can deny that modern weather forecasting, with its dependence on satellites, radar and global monitoring stations, has become a highly complex matter. Although many gardeners will be quite content to rely on broadcast indications of impending rain, drought, gales or frost – and some may have a thermometer and barometer to give more local guidance – an understanding of what is happening to the weather can add greatly to an appreciation of garden ecology. I have given some information above for those wishing to go right to basic principles, actually keep their own weather records and thus build up a fund of personal knowledge. For others, less technically inclined, a major virtue would be the ability to understand that basic medium of weather information, the weather map. Examples are shown here of the newspaper or television-style weather maps illustrative of four weather types of special interest to gardeners. These show the

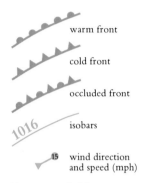

B blue sky
BC blue sky and cloud
C cloudy
O overcast
F fog
D drizzle
R rain

warm front

cold front

occluded front

_1016_ isobars

15 wind direction
and speed (mph)

Temperature in °C

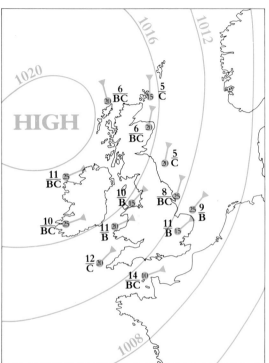

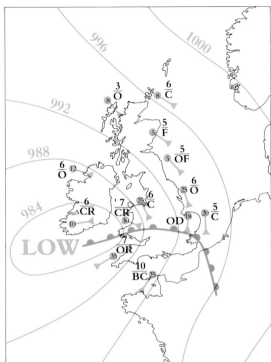

Noon 10 May
Late frost

An anticyclone is an area of high pressure surrounded by low. The isobars are wide apart, indicating light winds. Rain falls seldom and, although anticyclones can be associated with clouds, they usually occur with clear skies. A clear sky anticyclone in May, shown here, with its attendant night-time re-radiation of heat into the atmosphere is the most important of all garden weather signs. The anticyclone depicted is centred off the north coast of Ireland and, as winds move clockwise in such a system, most of the British Isles will experience northerlies. With the light winds, the clear conditions and low night temperatures will probably persist for a few days. Frames and greenhouses must be closed at night, tender plants in pots brought indoors and, if possible, some protection given to tender blossoming trees.

Noon 19 March
Likelihood of rain

A low pressure system or depression is centred over southern Ireland. The winds blow anti-clockwise around it and as the fairly close isobars indicate, winds are moderate. Warm, moist air from the south has risen over colder, northerly air and a so-called front has formed, clouds develop and rain will fall. (In practice, a warm front forms at the forward boundary of the rising warm air and a cold one at the rearward.) A belt of fairly general rain will pass northwards associated with the warm front, followed by bright intervals and showers associated with the more broken clouds of the subsequent cold front. This is good transplanting weather.

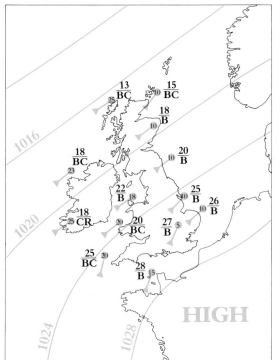

Noon 22 July
Settled summer weather
Like the left-hand map on p. 34, this too illustrates a high pressure system; but this one is in summer when, despite the clear skies, heat loss will never be great enough to cause frost. Winds again are light but, as this anticyclone is centred to the south of the country, they are southerly or south-westerly. There should be fine, settled weather although the north and west may have light rain as the winds pick up moisture over the ocean and clouds form over high ground. Watering is necessary over much of the country, greenhouses should be well ventilated and the benefits of mulching will be apparent.

Noon 6 January
High winds
High winds are indicated by closely spaced isobars. In Britain, they are often associated with a secondary depression, as shown here. The main depression is centred off the map to the south of Iceland and a secondary with its associated front has formed off the Western Isles, indicated by the bends in the isobars. This has increased the winds on its southern side, farthest from the main depression, and has its own wind circulation, apart from the main one. Easterly winds on the north side are less strong than the westerly winds on the south and this indicates first, southerly, then westerly gales as the secondary depression tracks rapidly in a north-easterly direction. In winter, the associated rain will probably fall as snow on high ground. Now is the time to hope that fence posts and gates were well secured in the autumn. Fruit cages and greenhouse glass will be particularly vulnerable to damage.

Cirrus
The high altitude wisps of cirrus clouds, scarcely visible against a clear sky are often thought of as fair-weather clouds although they may presage the approach of a depression. This underlies the significance of observing sequences rather than individual clouds.

Altocumulus
Fleecy bands of small, globular clouds at inter-mediate height are usually the indicators of periods of fine and settled weather.

Cumulus
The most familiar clouds of traditional British summer weather, cumulus have a straight base which indicates the height at which condensation of rising air masses takes place.

patterns most commonly associated with late spring frost, rain, settled weather in summer, and high winds; but it must be appreciated that the first three in particular are not exclusively associated with such patterns, and that it is seldom that the entire country experiences the same weather at the same time. Nonetheless, it is important to recognise the use of isobars (lines on a map linking points with equal barometric pressure) and other symbols and pictorial representations of depressions and anticyclones, fronts and wind directions.

The significance of cloud types

Clouds are masses of tiny particles of water or ice, floating at varying heights above the earth, and they are the most obvious visual indicators of at least some aspects of the weather prevailing at any given time. While the archetypal rustic may be able to gaze heavenwards and tell at a glance the weather patterns for the next few weeks, the more realistic and perhaps less romantic among us can still discover a good deal, if only we know what to look for.

Although the variations in the patterns of clouds may seem infinite, there are relatively few basic types, and even the recognition of these broad categories can be informative. There are three cloud groupings based on height: low clouds with a base not greater than 2,500 m; middle clouds with a base between 2,500 and 6,000 m; and high clouds with a base between 6,000 and 12,000 m. Clouds are also grouped on the basis of their general form and appearance into: feathery cirrus clouds, layer or sheet-like stratus, and heaped or globular cumulus. Many clouds are intermediate in type and are described by composite names; stratocumulus, cirrostratus and so forth, sometimes with the prefix 'alto' to indicate considerable height (altostratus, for example), or with the addition of 'nimbo' or 'nimbus' (as in cumulonimbus), indicating a 'heavy', rain-bearing cloud.

The sequence of cloud types over a period of time can be particularly revealing. For instance, during the passage of a depression, first cirrus and cirrocumulus are evident, followed by a thin covering of cirrostratus and then a thicker one of altocumulus. With the approach of the warm front, stratus and rainy nimbostratus blanket the sky. Subsequently, the arrival of the cold front brings cold air beneath warm, and strong vertical cumulonimbus development may occur with heavy showers instead of continuous rain. Later still, the cumulus gradually breaks up as the sky clears.

Cirrus
When the wind at high altitude is strong, the clouds are blown into tails or plumes.

Stratocumulus
A fairly dark, low altitude cloud that blankets the sky, often for long periods in winter when the sun is invisible for days or weeks at a time. It is stratocumulus clouds that so often give winter weather its dreariness.

Cumulonimbus
Rain clouds; a summer thunderstorm threatened when the photograph was taken and the cloud mass gradually billowed upwards to around 5,000 m.

Making the most of rainfall

It is unlikely that, in any part of the British Isles, the growing of a particular type of garden plant is impossible just because the rainfall is too high. Conversely, there is probably no area where the annual rainfall is inadequate for any species, given that other climatic factors are suitable. Problems arise, nonetheless, because, as every gardener knows, rain never seems to be falling when it is needed and always seems to be falling when it is least wanted. The most marked discrepancies between the amounts of rainfall in the different seasons tend to occur between those high, western districts with the highest total rainfall (Diag. p. 38), and the drier parts in the east of England which, in general, have a more uniform monthly distribution (Diag. p. 38). It is naturally the 'growing season' rainfall that is of greatest importance to gardeners and it is within this spring–summer period that small monthly, weekly or diurnal fluctuations can be most significant. Superficially, a July rainfall of 58 mm, such as that at Cambridge, may seem quite adequate for gardens, but it may all have fallen during two thunderstorms on, say, the 8th and 22nd of the month. Technically drought has not occurred, but with high temperatures and high evaporation rates, these 58 mm of rain will have had little effect on plant growth. Bear in mind that one square metre of leaf cover may lose over 5 litres of water per day at such times, and this amount will need to be replaced in the soil. Yet that 58 mm represents only about 32 litres of rain per square metre *for the whole month.*

Gardeners must conserve water during the summer months, be it directly from the clouds or from the tap, and use it wisely. Rain falling onto the garden will ultimately drain away through the soil; the rapidity with which this happens, and hence the likelihood that it will remain long enough to be used by plants, depends primarily on the soil type and structure. These factors are considered in some detail in Chapter 3, together with ways in which the water retentiveness of the soil may be

Those parts of the British Isles, like Oban, that have a high overall annual rainfall tend to be those with the most marked seasonal fluctuations. Lower rainfall areas such as Cambridge have their total rain much more uniformly distributed through the months.

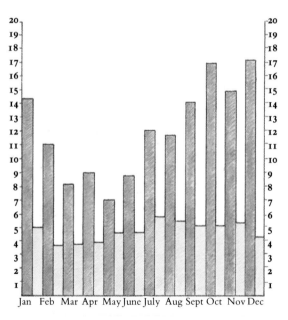

Monthly rainfall Oban (total 1451 mm)

Monthly rainfall Cambridge (total 588 mm)

altered. Rainfall may be collected before it reaches the soil, however, and used, with advantage, during periods of dry weather, especially if the use of hose-pipes is restricted. The obvious way of doing this is with a rain-water butt, collecting the water that drains from a greenhouse or shed roof. Modern butts are made of plastic, with taps and removable covers; the more attractive, traditional ones, from wood. It should be remembered, however, that rain-water butts harbour a variety of wild-life, including some organisms that cause plant disease. A useful maxim, therefore, is only to use such water outdoors, for, in the warmth of the greenhouse, any such organisms will be able to incubate and cause trouble. And rain-water shouldn't be stored in zinc-coated or iron vessels, as it will accumulate harmful metal residues.

Much has been written about the relative merits of rain and tap-water, but, except where lime-hating plants are being grown in an area with very hard (limey) water, the difference is not one that need concern gardeners. The science of crop irrigation is complex, but it is important to appreciate one simple, basic rule of watering, especially to avoid wasting resources at times of water shortage. If, in general, you ensure that a plant has most water at the time it is producing those parts that are the reason for its cultivation, then you will not go far wrong: flowers at flowering time, fruit trees and bushes as the fruit are swelling, and vegetables as the roots, heads, fruits or tubers are forming.

Making the most of sunshine – greenhouse principles and function

A greenhouse at its simplest is a translucent structure in which plants are grown and where the sun's heat is enhanced in some way. All green-houses, therefore, are 'heated' in the sense that infra-red and short-wavelength visible light components of the solar radiation strike them and pass through a translucent surface. Much of this radiation is re-emitted in the form of long wave-lengths which cannot pass back through the glass and are therefore trapped. Such a basic structure is popularly called an unheated greenhouse, to differentiate it from one in which some form of artificial heat is supplied in winter – either mini-mally, to maintain the temperature at just above freezing so that non-hardy plants may be kept over the winter, or more elaborately, to enable warm-climate species to be grown actively all year round. Many factors must be taken into account in greenhouse operations, but the most im-portant initial considerations are shape, siting and constructional materials.

Shape

Two main factors should be considered here: the influence that the shape has on the usable internal space, and the effect that the angle of the translucent surfaces has on light transmission. While time of day and season of the year are obvious variables, a sheet of glass at right angles to the sun's rays will always allow through approximately 90% of the light striking it, whereas at an angle of incidence of more than 40°, the percentage falls off rapidly (provided, of course, that the glass is clean – a moderate amount of dirt will impair transmission by up to 40%). Thus, the most efficient shape of all for a greenhouse is a dome; but this is scarcely practicable other than on a small, novelty scale. Next best is a square house with four roofs of equal size, or a house with a semi-circular roof and vertical ends. Although, ideally, such a house should be as high as it is wide, the design is made more practicable by flattening the profile and, by using plastic instead of glass, is perfectly feasible. The next preference is a structure with a straight roof and sides 60–70° from the horizontal, while least effective is the commonest shape – a straight roof and vertical sides.

A lean-to, built against a white-washed south-facing wall, rather than a free-standing structure, has much appeal. Less of the valuable open garden space is utilised and the warmth of the adjoining building can be capitalised upon. Unfortunately, the reflection of the strong south light by the wall does not fully compensate for the exclusion of the north light and, on these grounds, lean-to greenhouses have shortcomings, although their economy of ground area should not be forgotten.

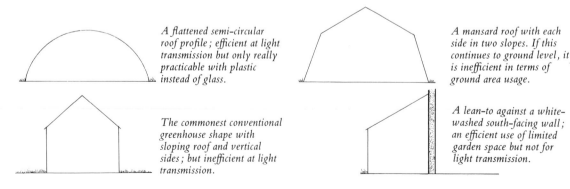

A flattened semi-circular roof profile; efficient at light transmission but only really practicable with plastic instead of glass.

A mansard roof with each side in two slopes. If this continues to ground level, it is inefficient in terms of ground area usage.

The commonest conventional greenhouse shape with sloping roof and vertical sides; but inefficient at light transmission.

A lean-to against a white-washed south-facing wall; an efficient use of limited garden space but not for light transmission.

Siting   Traditionally, greenhouses have been oriented north–south, but an east–west siting results in more uniform illumination and a more uniform transmission of light and use of solar radiation.

Wood or metal?   Wood is the traditional, 'natural' material for a greenhouse structure. Metal, in particular aluminium, is the modern alternative. Their relative pros and cons in terms of contribution towards the functioning of the greenhouse can easily be summarised: there are supposedly more nooks and crannies in a wooden structure to harbour pests and diseases; a timber frame is more flexible and less prone to wind damage; wooden glazing bars are thicker than aluminium and will cut out more light (but this is unlikely to be significant on a garden scale); and aluminium structures are more costly to heat artificially, for wood retains heat better than metal. There are also practical considerations like ease of erection, of alteration or of fitting attachments; and, of course, cost – both initially and subsequently. Whichever way you feel the balance is tipped, there is still another factor, for whatever else it may take account of, garden ecology, for me, is also about aesthetics. While a modern metal greenhouse may find its home in the modern garden of a modern house, it is as much out of place in the old garden of an old house as are aluminium window frames or glazed doors. I am unrepentant in my view that this is one instance where scientific reason should bow to aesthetics.

Glass or plastic?   Until recently, glass was the only possible material for a greenhouse. Clear plastics, especially polyethylene, are now viable alternatives, although it is important to appreciate the biological as well as the practical consequences of the differences between them. Both are similarly efficient at transmitting light of the wavelengths needed for plant growth (the 'photosynthetically active' radiation), but plastic is more efficient at transmitting the warming infra-red. This is a two-edged effect as the enhanced transmission works in both directions; hence, a plastic structure warms up more quickly, but cools down more quickly too. These comments apply particularly to new glass and new plastic, but plastic ages more quickly; it attracts dirt by static electricity and is degraded by ultra-violet radiation from the sun (some modern plastics less so), to become brittle, fragile and consequently prone to wind damage. Ultimately, I hope, aesthetic considerations will prevail here too when you make your choice.

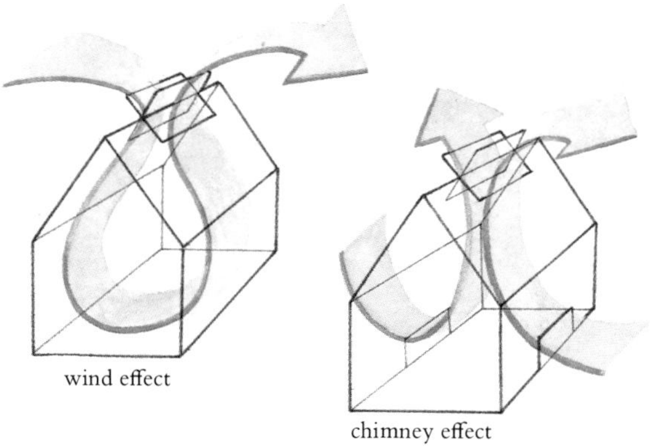

Ventilation

wind effect

chimney effect

Greenhouse ventilation is essential but usually inadequate. There are two main systems, each requiring a different ratio of vent to floor area (see text).

There are three functions of ventilation: to restrict the rises in air temperature as a result of the sun's heat, to restrict the rise in humidity due to water loss by transpiration from plants, and to admit air to supply the carbon dioxide needed for photosynthesis. In a small garden greenhouse, an ecological compromise is needed because heat-demanding plants like many succulents may well be grown alongside those such as lettuces needing cooler conditions, and humidity-loving cucumbers will accompany tomatoes which prefer a drier atmosphere.

An ideal pattern of vents for a small greenhouse is shown above, and before buying a greenhouse gardeners should look critically at its venting arrangements, especially in the case of a metal-framed type which would be more difficult to alter. (It is usually easier and cheaper to improve an inadequately vented greenhouse with louvred electric extractor fans rather than with more vents.) Automatic venting systems are helpful if the temperature range over which they can be set is sufficiently large, and if they are able to open the vents to an angle of at least 40°. Air circulation by the chimney effect method should use vents of area at least 15% of the total floor area, whereas without the side vents and using the wind effect method, the vent area needs to be as much as 30% of the floor area.

Shading

It is an irony of the British climate that, while a greenhouse magnifies the sun's warmth invaluably at certain critical times of the year, it over-magnifies it during the height of the summer. Ventilation alone is unlikely to counter-balance this in a small greenhouse, so some form of shading is needed as a control. Unfortunately, all materials that shade out the heating components of the solar radiation also shade out some of the light that plants need; and some actually change the spectral balance of the light, letting in too little of the important wave-lengths needed for photosynthesis. Far and away the most efficient method of cutting out some of the infra-red heating radiation, while having least effect on light quality, are some of the chemical 'paints' with which the outside of the greenhouse is treated. They are also much the cheapest. Even so, the inevitable ecological compromise will be needed if shade-loving plants such as ferns are in the same greenhouse as sun-loving cacti. A sort of solution can be achieved by placing different plants in different parts of the greenhouse (Diag. p. 43) and thus, in effect, mimicking the structure of a naturally layered plant community.

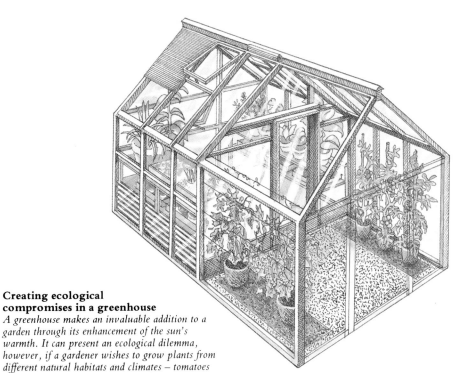

Creating ecological compromises in a greenhouse

A greenhouse makes an invaluable addition to a garden through its enhancement of the sun's warmth. It can present an ecological dilemma, however, if a gardener wishes to grow plants from different natural habitats and climates – tomatoes from hot dry areas and cucumbers from warm, wet ones, or alpines on the one hand and orchids on the other, for instance. An insulated internal partition helps in creating two different

environments, while the sunnier and shadier sides (and even the spaces under the staging) can be utilised to satisfy particular plants' requirements.

Supplementary greenhouse heating

Supplementary heating in winter increases the possible uses of a greenhouse; but it is important to decide the minimum temperature needed and then calculate the additional heating required. The graph shows the heat output needed in British thermal units per hour to maintain a temperature of $+7°C$ over a range of median minimal outdoor temperatures, which extend from about $-4.5°C$ in South-West England to about $-9.0°C$ in Central Wales or the Scottish Lowlands.

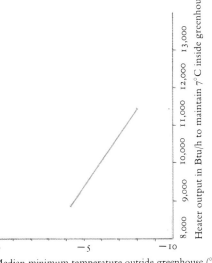

Supplementary greenhouse lighting

Supplementary lighting is neglected in garden greenhouses because the rewards very often do not repay the costs, although even a small amount of artificial lighting can improve the growth of the earliest seed sowings in a propagator. The commercial grower reaps massive benefits from supplementary lighting, however, and as the diagram shows it would be impossible without it to produce chrysanthemums at all from October to March, because the amount of radiation entering the greenhouse, even in southern England, is insufficient for them to grow satisfactorily.

Southern England

Northern England

Minimum quantity of radiation for chrysanthemums to grow satisfactorily

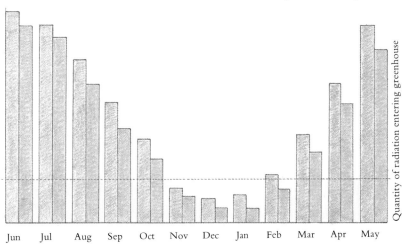

43

Windbreaks

In some gardens, existing shelter means that stakes and ties for tall growing plants are the only additions that the gardener needs to protect them from the effect of the wind. Nonetheless, in new gardens with only token boundaries, or gardens sited in exposed situations, the erection of windbreaks is a major priority.

The effectiveness of a wind-break depends upon its ability to reduce the speed of the wind, but not actually to stop it, for a solid structure deflects the wind upwards, resulting in a region of low pressure on the leeward side, with consequent downward air currents which may be more damaging than the original wind (Diag. p. 45). It has been found experimentally that a 50% permeable windbreak is the most efficient, and such a structure will reduce wind speed on the leeward side for a distance of 30 times its height – although the maximum benefit occurs for a distance of only 10 times its height. Thus, to protect a garden 20 m wide, a barrier 2 m high will be more than adequate. The next most important criterion in choosing a type of barrier should be aesthetic, although, regrettably, it is more likely to be economic.

Unless a garden is very large and very windy, the only windbreaks that should be necessary are those that also form the boundaries of the property, where the obvious choice lies between a wall, a hedge and a fence. A wall is far and away the most expensive per unit length and, in principle, conveys the inherent disadvantages of a solid structure outlined above. It is a good thing that most existing walled gardens are in towns where their effects on wind flow are unlikely to be significant, for the overall environment will already be fairly well sheltered. Indeed, gardening on walls themselves is a fascinating ecological exercise, and is discussed elsewhere in the book. The erection of walls round small gardens in open country is, however, to be viewed with caution; as boundaries, yes, but as wind-breaks, no.

Fences are available in a wide variety of styles. Again, however, where a wind-break is part of its function, the aesthetically pleasing structures

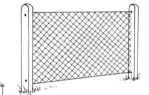

Chain link fencing provides a good windbreak but no privacy, and is ugly. Green plastic coated netting is best, and should be buried about 10 cm deep.

Overlapping softwood fence panels are attractive but impermeable to wind; they must be supported on very strong posts, with every second or third braced.

Hedges are ecologically and aesthetically excellent, but slow to develop. Beech is ideal for most purposes, but a fence screen may be needed while it matures.

Traditional paling fences are fine for small gardens. They have good wind permeability and are pleasing to the eye – but a nightmare to paint.

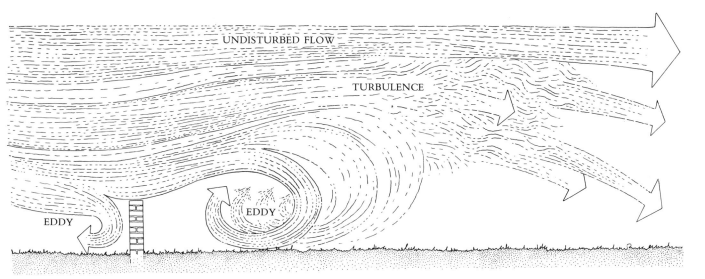

UNDISTURBED FLOW

TURBULENCE

EDDY

EDDY

An impermeable barrier creates turbulence and eddies on the leeward side which may themselves be strong enough to cause some damage.

of overlapping soft wood are of least value. Even if secure posts prevent them from being felled by winter gales, the lee-side turbulence may result, at the least, in most of the fallen leaves of the neighbourhood collecting in your garden. Some of the traditional slatted types that offer the desirable 50% wind permeability are much better. It should also be borne in mind that a growth of evergreen climbing plant over a slatted fence will reduce its permeability. Deciduous climbers present less of a problem: the winter permeability will be relatively unimpaired, and summer damage from strong winds is anyway unlikely.

Living wind-breaks of trees or hedges have many admirable features, both physical and ecological. The general role of hedges in garden ecology is considered in more detail in later chapters but, strictly as wind-breaks, they have much to commend them, provided they are not allowed to become too dense. Many of the attractive flowering deciduous hedging plants such as shrub roses have relatively little physical strength, whereas beech, though deciduous, retains its dead leaves, thus providing good, all-year-round effectiveness without the monotony associated with many conifers. On the sea coast, tolerance of salt-laden winds is the major factor to consider and, in mild areas, *Escallonia macrantha* makes an excellent choice.

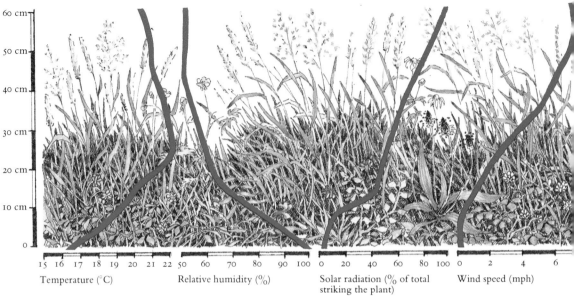

60 cm
50 cm
40 cm
30 cm
20 cm
10 cm
0

15 16 17 18 19 20 21 22	50 60 70 80 90 100	0 20 40 60 80 100	0 2 4 6
Temperature (°C)	Relative humidity (%)	Solar radiation (% of total striking the plant)	Wind speed (mph)

There are marked microclimatic changes close to the ground – temperature, solar radiation and wind speed fall off dramatically compared with the levels in the air above the vegetation, and humidity rises sharply. These have significance for the plants themselves and for the many other organisms dependent upon them.

Microclimate

The climate and weather facts presented in this chapter as maps, graphs and diagrams, relate to national patterns, and the data on which they are based are obtained from a country-wide network of recording stations. That there will be variations within a region is fairly obvious, and if your garden is several kilometres from the nearest official meteorological station, that is a very sound reason for building up your own fund of measurements. It is equally important to recognise the smaller but highly significant variations that exist within a garden, for while there will be uniform rainfall across the plot, some areas of soil will receive more than others, by virtue of the shelter present. Similarly, some parts of the garden will be better protected against wind, better sheltered from frost or more shaded from sunshine. All these factors are components of the microclimate and while most micrometeorological measurements are difficult to make, gardeners should be aware of the existence of wet and dry, cool and warm, or sunny and shady areas of their gardens. The subject is large and complex and only a few examples are given; they illustrate, however, that all may not be what it seems.

Most gardeners would comment that south or south-west facing walls and slopes are warmer than others. This is only partly true, for, in June and July in our latitudes, east or west facing walls receive more of the sun's warmth than do south facing ones; an indication of where

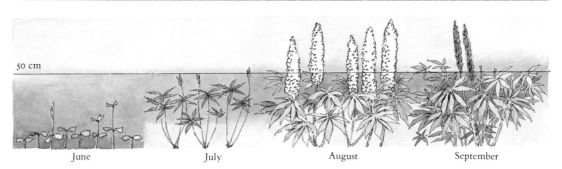

The changes in temperature in a flower bed as the plants grow during the summer. Early in the season, before the leaves are touching and the canopy closed, the bare soil heats up and re-radiates warmth close to the ground. This is a critical time for the tender young plants, which have to endure abnormally elevated temperatures. Later, the soil is covered, the temperature among the plants less harsh, and growth proceeds more rapidly. The maximum temperature is then near the top of the plant cover, where the horizontal upper leaves trap the sun's warmth.

150 cm

100 cm

50 cm

June   July   August   September

summer-flowering wall plants are best placed. South-east facing slopes are generally warmer than south-west facing ones; assuming, of course, that other microclimatic factors, such as exposure to wind, don't override this effect.

It will come as no surprise to learn that within a plant community such as a hedgerow or small copse, the climatic variations can be considerable. But it should also be remembered that within smaller communities (a bed of dahlias, a row of runner beans, even the grasses of a lawn turf) differences exist in temperature, light penetration, humidity and wind effects that will be of importance to the plants in question (Diag. p. 46). Such differences will themselves vary, not only with the gross macroclimatic variations during a season, but also because they will be influenced by the stage of growth reached by the plants concerned (Diag. p. 47).

For those beginning to despair of ever coming to grips with it all, worry not! Such facts are presented here, not because every gardener needs to appreciate in the most minute detail the microclimate of his garden, but simply because a general awareness that such phenomena exist will aid his understanding of garden ecology and, hence, the fulfilment he derives from his gardening. Some of this will, I hope, become more apparent in later chapters.

Chapter Three

Rocks, soil and the garden

With few exceptions, plants grow in soil. It is, in many ways, the very basis of gardens and, according to the popular gardening adage, 'the answer' lies within it. Given appropriate climatic factors, it is certainly the nature of the soil at a particular site that determines whether or not an individual type of plant will grow satisfactorily. But ask a gardener to define soil, to explain what it comprises and, even more pertinently, ask him to describe the differences between a range of soils, and the odds are high that you will be confronted with silence and bemusement. This is unfortunate, not only in that a passing acquaintance with soil biology, physics and chemistry will assist in the choice of plants and management of a particular site, but also because it will open up whole areas of interest that make for a greater appreciation of the environment and natural history at large.

It should be understood at the outset that soil is an extremely complex medium, immensely variable in constitution and, indeed, in behaviour under different environmental conditions. Nonetheless, all soils contain greater or lesser amounts of three types of component; the living, the once living but now dead, and the never living. Dig a flat-sided hole in any part of your garden and all three will be apparent. (Indeed, dig a hole sufficiently deep and you will probably see rock beneath your soil too.) In the side of your hole, or, to give it a technical name, soil profile, you will be able to see small creatures; earthworms, small beetles, insect larvae, together with plant roots. These represent the larger living components of the soil, although, as will be apparent later, the microscopic life is in many ways more significant for plant growth. Pull apart some of the soil and, depending on where your garden is situated, there will be a varying quantity of small pieces of partly decayed plant remains. These, again together with microscopic fragments, are the once-living but now dead fraction of the soil. Having pulled apart the soil your fingers will be moist, and this provides evidence for the presence of one of the never living components of the soil, the soil water. Look again at your soil, and you will see greater or lesser numbers of holes or spaces in it; in total, these may occupy up to 60% of the soil volume and contain the second of the non-living parts, the soil air. What remains comprises the vast bulk of the solid component of most soils, the soil mineral matter; it is important to know where it came from.

All soil minerals were once part of rocks, although they may well have passed through many stages of degradation, and even reconstitu-

ACID   NEUTRAL ALKALINE

3   4   5   6   7   8

The relationship between pH (indicated across the top) and various soil activities and nutrient availabilities – the width of the bands represents relative magnitude or activity.

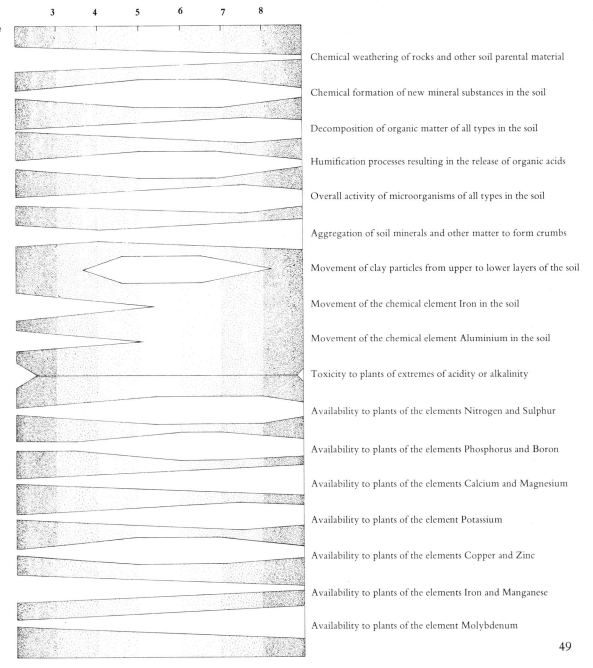

Chemical weathering of rocks and other soil parental material

Chemical formation of new mineral substances in the soil

Decomposition of organic matter of all types in the soil

Humification processes resulting in the release of organic acids

Overall activity of microorganisms of all types in the soil

Aggregation of soil minerals and other matter to form crumbs

Movement of clay particles from upper to lower layers of the soil

Movement of the chemical element Iron in the soil

Movement of the chemical element Aluminium in the soil

Toxicity to plants of extremes of acidity or alkalinity

Availability to plants of the elements Nitrogen and Sulphur

Availability to plants of the elements Phosphorus and Boron

Availability to plants of the elements Calcium and Magnesium

Availability to plants of the element Potassium

Availability to plants of the elements Copper and Zinc

Availability to plants of the elements Iron and Manganese

Availability to plants of the element Molybdenum

49

tion, before they arrived at their present-day condition. Moreover, while several different environmental forces may have worked on the soil to move it from the site of the rock with which it originated, it is apparent from a comparison of the geological and soil maps of the British Isles that their general patterns are the same. And, just as soils are immensely variable, so too are the rocks from which they are derived (the 'parent' rocks, as they are known). The British student of geology is, in fact, extremely fortunate, for there are probably no areas of the world of comparable size in which such a wide variety of rocks occurs; almost all the major periods of rock formation are represented here. These rocks can be grouped into three main divisions – igneous, metamorphic and sedimentary – on the basis of their modes of formation.

Igneous rocks are formed from molten material originating within the earth, either like granite at some depth beneath the crust, or like basalt during the course of volcanic activity. In some parts of the world they are of course, still actively being formed, when molten mineral matter is forced to the surface in volcanoes. Reassuringly, no igneous rocks have been formed in the British Isles for about 40 million years, but they are nonetheless responsible for much of the upland scenery of the north and west of the country; upland, because many of these rocks are extremely resistant to the eroding effects of wind, water and ice. Nonetheless, slowly, erode they do, and their constituent minerals are moved, by these same agencies of wind, water and ice, to be deposited elsewhere.

Over the course of geological time, these deposits may themselves be compressed together, harden and become rocks too. By virtue of their mode of formation, such rocks are known as sedimentary rocks. Sandstone is a common example of a sedimentary rock; but so are limestone and chalk, where the original minerals have not been deposited directly, but have been used by aquatic creatures to form shells, these in turn providing the raw material of the sedimentary rocks. Generally, sedimentary rocks are softer and more easily eroded than igneous and, where they occur as fairly high mountains, it is because geologically they are younger and have been exposed to erosion for a shorter time.

Both igneous and sedimentary rocks may be changed or metamorphosed structurally and chemically by the heat and pressure of earthquakes and earth foldings into the third group, the metamorphic rocks. Slates, marble and schists are examples of metamorphic rocks.

The similarity in the patterns of soil type and solid geology distribution is more apparent in some areas than others. It is shown particularly clearly in these sections of the 1: 63,630 Soil and Geological Survey maps of part of north Berkshire and south Oxfordshire

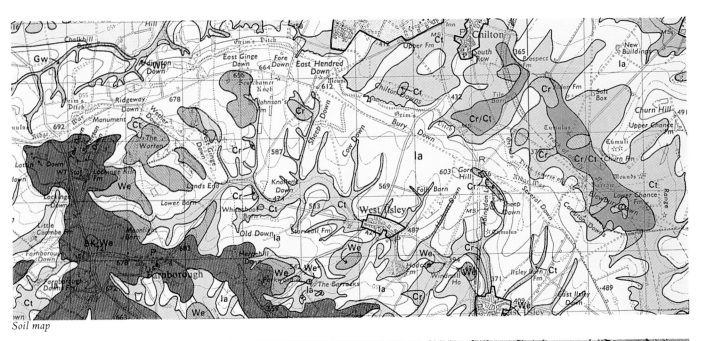

Soil map

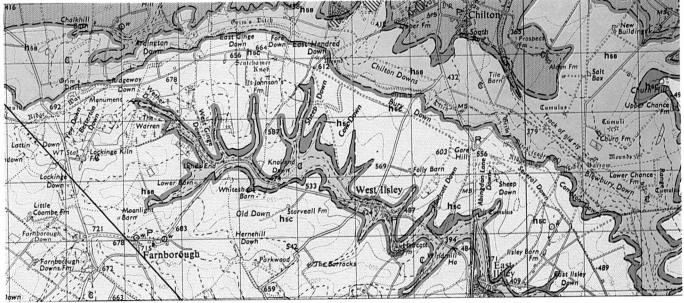

Geological map

From what has been said, it will be apparent that the type of rock in a particular area dictates very largely the nature of the landscape. It can also dictate the differences in gardening conditions; most of the variation for instance between a garden in the extreme south-west of England, and one with a very similar climate in the extreme south-west of Wales, is due to the presence beneath the former of granite and other acid igneous rocks, and beneath the latter sedimentary limestone, which have each given rise to a very different type of soil. Nonetheless, I remarked earlier that the soil in a particular garden or area needn't bear a close relationship to the rock lying beneath it – its parent rock may be many kilometres away – and nowhere is this more commonly true than in a river valley. Central London is a case in point. Look at a map of what is termed the solid geology of this area and you will see that the underlying 'rock' is, in fact, not a hard rock at all, but the infamous London Clay.

Close to the River Thames, however, this clay has had a fine, granular mineral material, termed alluvium, deposited on it by the river; indeed in the lower reaches of the river, around Woolwich for example, this alluvium deposit is several kilometres wide. It comprises mineral matter eroded and transported from rocks many kilometres upstream and contains substances different from those in the underlying clay. It is this river alluvium (known as a drift deposit), not the bed rock itself, that is the basis of soil close to the river. Similarly glaciers, wind and the sea can also bring about widespread transport of mineral raw materials. Thus, the growing conditions and types of plant that establish naturally, or may be cultivated, can vary widely within a small area if, for some reason, the soils are markedly different.

But how does rock, any rock, soft or hard, bed rock or alluvial deposit, become soil once it has stopped being blown or washed around the countryside? The process is both long and complex, and many interacting factors bear upon it. The chemistry of the mineral matter may itself be changed by the action of rain and by the eroding, digestive effects of the first, usually simple, forms of plant life that establish themselves on the crude material. Gradually, as these plants die, their remains are added to the mineral matter to form the first organic component of the soil and this, in turn, enables more (and usually more advanced) forms of plant life to grow. Their growth and nutritional processes, too, impose a chemical influence and on their demise they add further to the soil's make-up. As animal life establishes itself, the chemical changes are extended, and animals' mobility means that they can aid and accelerate the mixing brought about by wind and rainwater, further to blend the whole. It has been estimated, for example, that up to 100 tonnes of soil per hectare are moved by the earthworm population in the course of a single year.

If we return to the soil face in the side of the hole in your garden, we can see these processes of soil formation, 'frozen' in the layers of what is termed the soil profile. If you have dug down far enough, the base of the profile will reveal the bed rock. Immediately above it is a region of slightly changed and eroded mineral matter, but largely lacking any organic component. This is known popularly as the sub-soil and above it are layers, or horizons, that contain an increasing proportion of plant and animal remains, and that are increasingly different structurally and chemically from the rock itself. In a garden, of course, especially an old garden, the demarcations between these various layers will have become blurred as digging and general cultivation has mixed them together; a hole in a less altered environment will show the divisions more clearly.

Although I have stressed the great variability of soils in terms of their chemical and physical make-up, they are not so variable as to have defied the wit of man in classifying and naming them. The Soil Survey in Britain, in common with comparable government-sponsored surveys in other countries, has categorised soils in various ways. Soils with particular mineralogical, structural and other features are classified in Series and Associations, usually named after a locality where such soils typically occur. Underlying this scheme, however, is a categorisation with which gardeners will be more familiar, even if they are unaware that it does have a very precise meaning. The basis of the physical classification of soils is in the relative percentages of sand, silt and clay that they contain. The demarcation between these three types of particle is clearly defined by the following size limits:

Particle type	Particle diameter range (mm)
Sand	0.006–2.0
Silt	0.002–0.06
Clay	<0.002

The British countryside is enormously varied and, overall, the British Isles probably display a greater variety of plant communities and landscapes than could be found within an area of comparable size anywhere else in the world.

This variation is due very largely to the variation in soils, which reflect, in turn, the great variety of underlying rock. The acid heathland in West Sussex (top left), characterised by heathers, pines and birches, is formed on a soil with a pH around 4.0 overlying Cretaceous and Tertiary sands. By contrast, the grassland of the Suffolk coast (bottom left) lies over marine alluvial soils with a pH of 7.5, while the rowan Sorbus aucuparia (top right) is at home in the rugged countryside of Perthshire where the soil is almost non-existent and it grows from a cleft in streamside granite and schist boulders.

Many factors contribute to the formation of soil from rock, but few work as quickly or dramatically as the sea, which has eroded and shaped the chalk of the Dorset coast (bottom right).

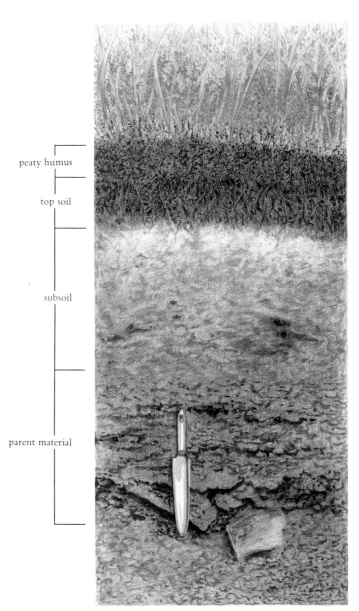

peaty humus

top soil

subsoil

parent material

The vertical face of the side of a hole dug into the soil reveals what is termed the soil profile. In a relatively undisturbed natural habitat, as opposed to the well mixed soil of a garden, the division of the soil into fairly clearly marked layers or horizons is apparent. The professional soil surveyor subdivides these horizons into detailed structural and textural categories, but the general divisions labelled here are adequate for gardening purposes. The relative depths of the different horizons vary markedly between soil types; the soil shown here is an upland forest soil. The trowel is 25 cm long, including its handle.

Once a gardener begins to dig and cultivate a soil, it becomes apparent that it possesses certain physical characteristics. And once he has worked with more than one type of soil, it will be apparent that these can differ widely. Some soils dry out very quickly; some become very sticky when wet, perhaps to the extent of being unworkable in winter; others very soon develop a hard crust or cap after heavy rain, or after being walked on. Such features have an important bearing both on the ease with which soils can be cultivated and on the types of plant that grow most satisfactorily in them. No-one gardens for very long before he discovers that carrots and other root crops grow best in a light, free draining soil with no large clods, or that a moisture retentive soil is essential for many of the most choice species of *Primula*. Two words are often used for the physical attributes of soil; and often incorrectly. These are texture and structure.

The texture of a soil is an expression of the relative proportions of sand, silt and clay particles that it contains. The way to perform a simple textural analysis of your soil is described on p. 63, and the rather curious triangle shows how, from these relative proportions, the soil can be named accurately as a sand, clay or loam. In common usage, it is sandy soils that are often described as light, and clay soils are said to be heavy.

The structure of the soil, on the other hand, describes the manner and extent of aggregation of the mineral particles into what are usually called crumbs. Within the crumbs, between the individual mineral particles, are holes or pores that contain the soil air and/or water. In a clay soil, these are much smaller than in a sandy soil. There are also pores, however, *between* the crumbs. It is this blend of large and small pores that makes for what gardeners call a well-structured soil. In contrast, a beach sand or a very heavy clay is virtually structureless.

There remains the third major soil component to consider, the once living but now dead organic matter or, as it is often called in gardening, the humus. This too contributes significantly to soil texture. Although soil scientists classify soils, in terms of organic matter, by relating the percentage of organic carbon they contain to the percentage of clay in the mineral fraction, it is sufficient for gardening purposes to appreciate that most soils contain between 1 and 5% organic matter, although in a peaty soil, it will be very much higher.

Clearly both the texture and structure of a soil have major bearings on its drainage characteristics. Water will drain much more rapidly

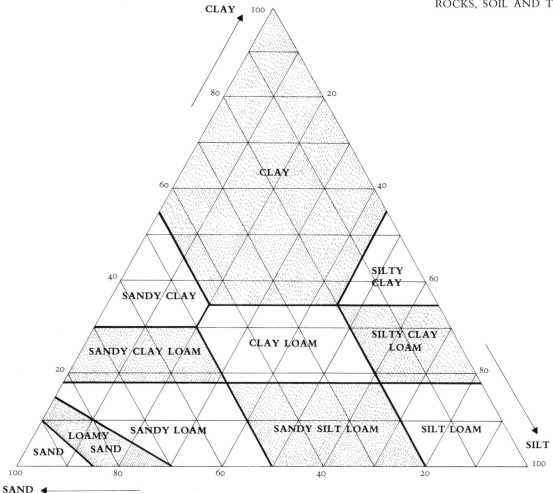

Given a basic textural analysis of soil into the relative percentages of sand, silt and clay particles that it contains, it is possible from this diagram to place the soil in one of a number of broad categories. Further detailed analysis such as a soil laboratory might perform will enable the soil to be still more precisely defined.

through a sand, having poor structure and large pores, than it will through a close-packed but also poorly structured clay. Drainage, or more precisely water retentiveness, can be altered, moreover, by the relative amounts of organic matter or humus in a soil, for this tends to act like a sponge, soaking up and holding the moisture.

All these features of the physics of the soil do therefore have important ecological significance: plants requiring abundant and constant moisture around their roots, for instance, will grow in very clayey or organic soils, rather than in sands. Nonetheless, the physics of the soil generally takes

Alkaline soil
Good indicator plants for alkalinity are less obvious than those for acid soil, but clematis are reliable.

Acid soil
The presence in an area of azaleas and rhododendrons, camellias and kalmias indicates an acid soil.

second place to the soil chemistry in ecological significance, and to one aspect of soil chemistry in particular: its relative acidity or alkalinity, usually defined as its pH. Technically, pH is a rather complex unit of chemical measurement; but for present purposes it is sufficient to know that it is measured on a scale ranging from 1 (very acid), to 14 (very alkaline), the mid-point of 7 being known as neutral. In practice, British soils fall between the 3.5 of a very acid peaty soil and the 8 of a thin soil overlying chalk rock. The difference in soil pH is the greatest single factor in accounting for the natural presence of certain plants on some sites and their absence from others with otherwise similar environmental attributes. And, in gardens, the pH is the most important single soil factor dictating which plants will or will not grow satisfactorily. Its effects are mediated in several ways, but of greatest importance is that most elemental plant nutrients are taken up by plants less readily from an alkaline soil. Although they may be present in adequate quantities, the high pH renders them unavailable. Plants such as clematis, and some gentians, that can tolerate this handicap and grow satisfactorily in alkaline soils, are known as calcicoles; those such as many vegetables and, indeed, other species of gentian, that cannot, are called calcifuges. A group that includes rhododendrons and many heathers are intolerant of all except the most acid soils.

The validity of my comment at the beginning of this chapter, that soil is a highly complex medium, will, I hope, now be evident. Soil is, moreover, a dynamic and ever-changing medium, for the many chemical, physical and biological influences on it ensure that, little by little, it too evolves. Gardening accelerates this process in some ways, but retards it in others. Perhaps more than anything, gardening depletes a soil of its organic component, and with it, of course, much of the nutrition that plants require. Especially when growing vegetables, we leave plants in the soil only until they have fulfilled the role we have allotted for them; we never allow them to die and decay to the state where some of their constituent materials are returned to the soil. And we cultivate fast growing, specially bred and alien species that exhaust the nutrient resources more rapidly than any native British woodland plant ever did. But part of gardening, good gardening, is in making up for these abnormal constraints that we place upon our soil. Most obviously, we do this by adding nutrients in the form of fertilisers, and organic matter in the form of manures and composts (see p. 64). We can make amends,

The symptoms of clubroot disease are familiar to every vegetable gardener. It is rarer on related wild plants; the different soil environment and a lack of resistance in cultivated plants may partly explain this.

too, for our unreasonable demands on the soil ecology by the expedient of rotation. Because different types of plant have different nutritional requirements, the whole spectrum of soil nutrient will never be tapped if we constantly grow only one type of plant on the same area of soil. While this cannot be helped if you have planted slow-growing perennials, it can and should be avoided with quick-growing and rapidly-maturing vegetable crops which will themselves become even more dependent on abnormally high levels of artificial feeding. By changing regularly the types of vegetable grown on any particular area, the need for artificial feeding will be kept to a minimum and the soil's natural resources will be tapped to the full.

Nor should we forget the need to counteract such physical effects as soil compaction; in particular, the compaction brought about by you and I walking or, worse, driving a cultivator or lawnmower over it repeatedly. A compacted soil surface impedes the free penetration of air and water, and of fertilisers too. It hinders seed germination and encourages the growth of moss and algae which will themselves, in turn, help to deplete the soil of nutrients.

It will now be apparent that, in the soil, as in other components of the environment, gardening disturbs what is popularly known as the balance of nature. A diagram (p. 17) can show vividly how complex are the natural interactions between plants, animals and nutrients, but our activities change this essentially cyclic process into a much more linear one. Apart from the basic mineral raw materials derived from the parent rock and the substances present in the soil organic matter (themselves derived originally, of course, from the soil, or manufactured by plants during the process of photosynthesis and constantly replenished as they die and decay), there is one other important source of 'new' soil nutrient. Plants like peas, beans and lupins that belong to the family Leguminosae, bear on their roots small swellings known as root nodules. These contain specially adapted bacteria, mostly of the genus *Rhizobium*, which possess the unusual and important attribute of being able to convert nitrogen gas from air (the soil air, to be precise) into nitrogen-containing chemicals on which plants can feed directly. The value of such plants in the environment as a whole, and particularly in a crop rotation scheme, will immediately be apparent.

The root nodule bacteria are but one specialised example of the multitudes of micro-organisms that inhabit the soil. The breakdown and change of one soil chemical to another is brought about only partly by the action of the physical elements. Also essential to the process are bacteria and soil-inhabiting fungi, aided by small forms of animal life. Such organisms are as effective a part of any environment as the larger ones that we see. Clearing a woodland to form a garden will have an immediate and obvious ecological impact on any oak trees or bluebells that once grew there, or on the squirrels and voles that fed on them, but the impact on the soil micro-organisms will almost certainly be overlooked. Change the macro-environment, however, and you change the micro-environment too, with the result that the biological functioning of the soil may be impaired and, what is more immediately apparent, less beneficial micro-organisms may come to the fore.

Remove a plant (call it a weed if you will) that grows naturally on a site and perhaps you remove a species of mite that lives on its roots. The absence of this mite could mean that there is nothing to feed on the spores of a microscopic species of soil fungus, that just happens to attack the tissues of certain members of the plant family Compositae. Try then to grow lettuces on your new garden carved from the oak wood, and the soil fungus will play havoc with your plants.

Although firm evidence is lacking for many of the micro-ecological effects of gardening on the soil, the condition known as clubroot is perhaps indicative of what may happen. This disease, caused by a soil-inhabiting fungus, devastates plants of the horticulturally very important family Cruciferae and, as a consequence, is one of the biggest problems for the grower of brassicas, especially on the slightly acid soils where the fungus thrives. Why is this organism not equally devastating to wild plants of the same family? Is it because the natural habitat of the cabbage is the alkaline soil of the sea cliff and that we have imposed, in our vegetable gardens, soil conditions much more favourable to the pathogen? Or have we, by removing the plant from its native environment, perhaps removed some natural soil-inhabiting predator of the clubroot organism? If either, or even neither, of these conjectures is correct, there is no doubt that we have taken a plant and a pathogen from one environment to another; so we shouldn't be surprised if their interactions in garden soil are somewhat different from what happened in the wild state.

Top *Leatherjackets, the larvae of crane flies are among the commonest of the many pests that spend at least a part of their lives in the soil.*

Bottom *The roots of all members of the family Leguminosae bear small swellings. These are caused by bacteria able to convert atmospheric nitrogen into forms that plants can use directly.*

Managing the soil

Discovering what is present already – soil analysis

While 'managing the weather' is essentially a matter of coming to terms with it, of enhancing some of its effects, and, if possible, minimising others, the soil is considerably more amenable to the gardener's influence. To some extent, it is possible actually to change features of the physics, chemistry and indeed biology of your garden soil by appropriate amendments. The very experienced gardener may be able to judge a good deal of what is required simply by looking at the soil, by feeling it and by observing the types of plants growing in the garden and how well they perform; but for the rest it is always sensible to undertake a methodical appraisal to ensure that time and money are spent most effectively. The professional soil analyst will make comprehensive measurements and descriptions of a wide range of features, in order to assess the outcome of any possible future use for the land. But gardening requires very much more basic details, and the following simple scheme will yield valuable information in return for the minimum effort.

It should be remembered, moreover, that valuable intelligence can be gathered by examining the types of wild plant growing in the garden (see Chapter 5), and on the symptoms displayed by different types of cultivated plant. Remember, too, that information about the soil should always be considered in relation to the local climatic pattern, partly because plants will respond to both soil and climate, and partly because climate has some direct effects on the soil itself: a wet, heavy soil in a very cold area may well be frozen and unworkable for a large part of the winter, while a light soil in an area of low rainfall may present serious problems of water shortage.

Physical analysis of soil

Although the different horizons in a soil are interesting, there is no need to analyse the make-up of the soil profile in any great detail. Nonetheless, digging a hole is useful in revealing how deep the soil is and at what depth you are likely to reach the bed rock; where the soil lies over a drift deposit, it is quite possible that you will need to dig a very deep hole indeed before you strike resistance. The soil itself will contain greater or lesser numbers of stones, and it is useful to see if you can learn anything from their nature. Although specific geological identification of many stones is difficult, the presence of pieces of limestone or chalk will have special significance when possible soil amendment is considered. Look also for the presence of a hard, crusty horizontal layer within the soil. Such a layer is called a pan, and results from the deposition of mineral matter washed down from nearer to the surface. Its existence can mean that drainage and root growth may be impeded even on an apparently light, free-draining soil; and deep digging will be necessary to break through it.

Bear in mind that most cultivations take place within the top 45 cm of soil and, indeed, samples for analysis are best taken from within the upper 15 cm. By taking samples from the sides of a hole, rather than scraping them from the soil surface, you will minimise the likelihood of contaminating soil from elsewhere affecting your tests. A first general indication of soil texture can be learned by rubbing a small amount of moistened soil between finger and thumb, when its predominant component of sand, silt or clay, will be revealed by relative sharpness, smoothness or stickiness respectively. More precisely, the proportions of these components can be demonstrated by shaking up a trowelful with about a litre of water in a tall, preferably flat-sided, glass vessel and then allowing it to settle. The sand, silt and clay will sink and form layers, with the sand particles at the bottom. Organic matter in the soil will float to the surface.

Changing your soil – physical considerations

Although it is perfectly possible to take literally the notion of changing your soil by importing fresh top-soil from elsewhere, this is not an exercise to be undertaken lightly. If you have acquired a garden that represents little more than a building site, from which the top-soil has been removed (and quite possibly sold to someone else to alleviate *their* problems), buying a quantity of good quality loam is certainly the quickest way to provide yourself with at least a semblance of beds and borders. Nonetheless, bear in mind that a cubic metre of soil weighs approximately 1 tonne and that, as you need at least 30 cm depth to garden satisfactorily, this 1 tonne will cover an area of only about 9 square metres. The moral, if you are having a new house built, is to teach your builder some basic ecology, explain to him that top-soil takes many hundreds, if not thousands of years to form and that plants do not thrive in sub-soil. Try to ensure therefore that as the top-soil is removed from the site, it is stacked carefully, and then replaced when construction is complete. In 'normal' gardening operations, I believe that a gardener should only find it necessary to buy soil if he wants to introduce significant changes in the levels in his garden or to re-soil an old rose bed prior to replanting.

Soil texture can, of course, only be changed by altering artificially the proportions of sand, clay and silt. The addition of clay to a sandy soil, or vice versa, were once more popular than they are now; indeed, marling (the adding of alkaline clay to light, free-draining soils) was once a regular feature of our agriculture; in some areas old farmland is littered with the shallow depressions, called marl-pits, from which the clay was dug. It is a lengthy and laborious operation, only worthwhile if there is a good, local supply of the appropriate material (a state of affairs, incidentally, that underlies how dramatically soils can vary within a small area); otherwise efforts are better directed towards improving soil structure.

Proprietary soil 'improvers' designed to achieve this sought-after change in soil structure have been the subject of a minor growth industry in recent years. All of the materials added to soil to change its structure fall, nonetheless, into two main groups. They are either predominantly composts or manures, or they comprise a natural or artificial adhesive that binds together soil particles and so increases the number and size of soil crumbs.

Consider first the various unamended composts and manures. These will improve soil structure by increasing its water retentiveness, by opening up the compacted structure of a clay, and by helping to bind together sand particles and thus enhance crumb formation. They will also, to a small and valuable, but often overestimated degree, add nutrient to the soil (see Table 1). There is little to choose between garden compost, farmyard manure, peat, pulverised bark, spent hops or spent mushroom compost in respect of their contribution to structure, although the peculiar features of each should be borne in mind.

Garden compost must be well made, well decomposed, and must not contain diseased plant material nor abnormally high proportions of any one type of constituent (lawn mowings, for instance). Strawy farmyard manure should only be dug into the soil in the autumn for it will inevitably begin to decompose further *in situ*, and the micro-organisms responsible for this decomposition will deplete the soil of nitrogen. (Plants do not need much nitrogen in autumn.) It may also contain herbicide residues. Peat, in most areas, is likely to be rather expensive; pulverised bark is even more so; and neither adds any nutrients to the soil. Home-made shredded bark is not recommended, for it could contain toxic substances that inhibit plant growth. Mushroom compost may

Table 1   Nutrient content of organic manures and composts

Material*	Approx. % contents of main nutrients			Noteworthy features
Manures	N	P	K	
Strawy cow	1.2	0.4	0.5	Valuable as a soil conditioner in autumn only
Old cow	0.6	0.4	0.4	May be applied in early spring as less likely to deplete N
Stable	0.7	0.5	0.6	Lower moisture content than cow; easier to handle
Pig	0.6	0.6	0.4	Strong smell mitigates against it other than in rural area
Sheep	0.8	0.5	0.4	Low in moisture; easy to handle but rarely in large amount
Rabbit	0.5	1.2	0.5	Rarely available in useful amounts but useful P content
Chicken	2.0	1.8	1.0	Care needed to avoid N:K imbalance
Pigeon	3.4	1.4	1.2	Even higher N in relation to K than in chicken
Dog	0.4	0.2	0.1	Kennel waste should be composted first for health reasons
Composts				
Bracken	2.0	0.2	0.5	Better as a mulch than manure as may deplete N as it rots
Garden compost	0.7	0.4	0.4	Care needed to avoid weed problems if inefficiently made
Leaf mould	0.4	0.2	0.3	Rots very slowly; better made separately for mulching
Mushroom compost	0.6	0.5	0.9	High lime content; possible insecticide residues
Sawdust	0.2	<0.1	<0.1	Not recommended; rots very slowly and may be toxic
Seaweed	0.6	0.3	1.0	High in trace elements; attracts flies and best composted
Silage waste	0.2	0.1	0.6	Only useful for its potash content
Soot	3.6	0.1	0.1	Should be composted first; possible toxic residues
Spent hops	1.1	0.3	0.1	Care needed to avoid N:K imbalance
Straw	0.5	0.2	0.9	As bracken; compost first unless used as mulch

* *It should be appreciated that attributes of manures will vary with the types of bedding used for animals and with other differences in the conditions in which they have been kept. Nutrient contents will also vary with the age of the material and the conditions in which it has been stored.*

N = Nitrogen
P = Phosphate
K = Potash

contain a large amount of lime, and possibly insecticide residues. It is important, finally, to remember that the addition of organic matter to the soil cannot be a once and for all exercise. Once incorporated, it will continue its decomposition and must, therefore, be replenished annually.

Some of the proprietary soil improvers are little more than dried manure (sometimes poultry manure which commonly lacks straw or other bulky material and is, therefore, less effective as an improver of structure). Other materials contain greater or lesser amounts of added chemical fertiliser: although useful for small gardens in inner cities where natural supplies of organic matter may be scarce, they usually represent a very expensive way of buying a given weight of manure and fertiliser.

The second group of soil improvers, sometimes called soil conditioners (although this expression has no precise meaning) comprises certain chemical compounds of fairly well defined function. Products derived from seaweed tend to be rich in substances called alginates which, quite literally (but to an arguable extent) stick soil particles together and so improve crumb formation. (These seaweed products are also rich in plant nutrients, especially minor or trace nutrients, which some believe to be especially beneficial for plant growth; see p. 72). More recently, certain synthetic chemical compounds have shown great promise in achieving better crumb formation but have not, as yet, achieved widespread practical application.

There is a further feature of soil structure management to consider, and it relates to soil water. While the incorporation of organic matter will benefit, paradoxically, both drainage and moisture retentiveness, it is worth considering the soil surface of a natural plant community. The floor of a woodland, and indeed the soil under a meadow, is covered with a layer of organic matter: leaves and other plant debris in the process of slowly decomposing, but not yet incorporated within the soil by earthworms and other agencies. This covering is known as a mulch, and it serves the very valuable function of minimising the loss of water by evaporation through the soil surface. It also provides a measure of insulation against heat loss and thus lessens the likelihood of radiation frost damage. Most situations in gardens do not assist organic matter to accumulate naturally on the soil surface, and so there is every justification for adding it artificially to beds and borders. Certain cautionary points must be made, nonetheless, with regard to mulching. The operation does

And so to beds . . .
Gardeners normally grow plants in rows (left), with wider gaps between the rows than between the individual plants within the row. At first sight, the 'rows' in the commercial field (below) may seem similar, but they are grouped together to form beds, within which the plants are spaced equidistantly: the distance between the rows is the same as that between the individual plants. Thus each plant has more or less equal competition from all sides, instead of from close neighbours within the row and weeds between rows.

not, in itself, render soil moist; it helps retain the soil in the condition prevailing when the mulch was applied. Mulch a dry soil, therefore, and it will remain dry. Remember, too, not to mulch with undecomposed plant material; although lawn mowings are often used, they will begin to rot through microbial action *in situ* and in doing so will deplete the soil of nitrogen. Moreover, the very high temperatures generated during the decomposition process may cause damage to tender plants. In a natural community, only dead and dying plant material falls to the soil surface. Gardeners should remember this.

All the procedures outlined above depend on the addition of substances to the soil to improve its structure. Structure can also sometimes be helped, however, by the removal of a particular substance which is present in excess. A waterlogged soil is benefited most by the provision of a drainage system; and it should be remembered that even a garden with a highly improved soil will still give rise to serious problems if it lies over a poorly draining sub-soil, below the depth of normal cultivation. Drainage systems are not always easy to instal in gardens, largely because of the difficulty of siting a soakaway. Nonetheless, it should never be forgotten that purpose-built drainage systems, synthetic soil conditioners and abnormally large quantities of organic matter are not natural features of soils and, if all else fails, it is perfectly possible to grow plants on a poorly structured soil; you will simply be restricted to the types of plant for which this particular environment is the natural ecological niche. The background to this is discussed in Chapter 5.

On p. 60 I mentioned the ecological implications of compaction of the soil surface and, in gardens, this arises most often when the soil of beds and borders is walked on. Few natural plant communities suffer compaction to anything approaching this extent, and most are protected by the surface mulch of organic matter from the effects of rain, the most important natural compacting agency. Quite apart from its direct effects on plants and soil, a compacted surface means that the soil must be dug at least annually to permit rain, air and nutrients to enter. Nowhere is this more true than in the vegetable garden where the continual walking between the rows of plants can result, especially on a clay soil, in a surface texture akin to concrete. The soil structure will benefit enormously from adoption of the deep bed system, in which the soil is dug thoroughly, and organic matter incorporated, to twice the normal depth. Instead of conventional rows, the crops are grown in beds, approximately 1.2

metres wide and all cultivation is performed from pathways between them. Thus, there is no necessity to walk on the soil of the bed at all, and digging should be reduced to a once in five years exercise.

The ecological features of various ways of growing and spacing vegetables are discussed in Chapter 6.

Chemical analysis of soil

The professional soil analysis laboratory is an impressive enterprise using highly sophisticated and exceedingly expensive equipment. The identification and quantification of soil chemicals is a complicated matter, made more so because of their existence in the presence of, and sometimes in combination with, other chemicals that confuse the analytical procedures. While such precision is not needed for a garden soil, there is an important lesson to be learned from it. Treat with the greatest circumspection the claims made by manufacturers of small soil test kits sold for garden use. They will give you a crude indication of the relative amounts in the soil of some of the major nutrients needed by plants, especially phosphorus and potassium. They will be very misleading, however, in their analysis of soil nitrogen, which is so greatly influenced by the prevailing rainfall, temperature and other conditions that it can show fluctuating levels on an hourly basis. (The level of nitrogen is more meaningfully indicated by a novel method that involves the testing of small amounts of plant sap.) In fact, in many respects, the nutrient status of an unknown soil is best indicated by examining the symptoms of plants grown in it. In Table 2 are listed some of the most reliable plant indicators of deficiencies in the soil of the important plant nutrients.

Where soil test kits are especially valuable is in their use for checking soil pH, and they are more accurate than the small meters sold for this purpose, which actually measure not pH at all but a small electrical current set up in the soil between two dissimilar metals. Small strips of pH test paper, incorporating several different reagents, are even more convenient to use; unfortunately they are not as widely available to gardeners as they should be. The reliability of all of these methods is ultimately dependent on the care with which the soil has been sampled: the need is for at least five, and preferably ten, representative samples per 100 m^2.

The optimum soil pH for most common garden plants is about 6.5, and on p. 70 are some guidelines for ways in which it can be attained.

Comparison of the results from a popular garden pH testing kit with those from professional equipment

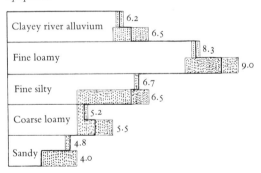

Clayey river alluvium	6.2 / 6.5
Fine loamy	8.3 / 9.0
Fine silty	6.7 / 6.5
Coarse loamy	5.2 / 5.5
Sandy	4.8 / 4.0

Five readings were made in water on samples collected from between 5 and 15 cm below the soil surface. The garden kit (bottom figure of each pair) did not permit accuracy greater than ± 0.5 units. pH testing papers impregnated with litmus or other indicator chemical are much cheaper and generally more accurate.

Changing your soil –
chemical considerations

I have already discussed the importance of soil pH, and touched upon the way in which its influence on nutrient uptake by plants is so important in determining which species grow satisfactorily on a particular site. Two questions concerning soil management present themselves; how can soil pH be changed? – and what other factors influence the nutritional status of the soil and its ability to satisfy plants' needs?

There is no disguising the fact that while the pH of soil can be increased without too much hardship, it is lowered (usually the preferred course of action) with infinitely greater difficulty. A soil with high pH is an alkaline soil; to change a soil of low pH to this state, simply add an alkaline chemical compound therefore. A group of compounds known under the generic name of 'lime' are widely and inexpensively available.

Table 2   Useful indicator plants for soil nutrient deficiencies

Nutrient Element	Indicator plant(s)	Symptoms of deficiency
Nitrogen	Brassicas	Plants small and with purple, red or yellow foliage tints
Phosphorus	Currants	Dull bronzing of the leaves with brownish spots
Potassium	Tomatoes	Undersized fruit, often with non-uniform ripening and some scorching of leaf margins
Calcium	Tomatoes	'Blossom end rot' (dark lesions at blossom end of fruit)
Magnesium	Lettuces	Yellowish marbling on the leaves
Iron	Raspberries	Pale leaves with dark green veins
Manganese	Peas	'Marsh Spot' (dark lesions in cotyledons when seeds are pulled apart)
Boron	Sweet Corn	Pale stripes on leaves and cobs fail to form properly
Molybdenum	Cauliflowers	'Whip-tail' (leaf blade is markedly narrowed)

Add lime to an acid soil, and the pH will be raised. Certain facts should be remembered, nonetheless; it is essential to conduct a pH test, to consider the types of plant that you wish to grow, and if possible to examine how well they grow at present, and thus determine whether your soil actually needs liming at all. Gardeners seem to fall into two camps; those who never add lime to their gardens and those who do it to excess.

Given the need to add lime, the choice of material to use lies between calcium oxide (burnt lime or quicklime), calcium hydroxide (slaked lime), calcium carbonate (ground limestone or ground chalk – the latter is simply a chemically purer form), and the mixture of calcium and magnesium carbonates known as magnesian limestone. Calcium carbonate is the cheapest and calcium oxide the most expensive form of lime, and they are respectively, the least and most efficient at neutralising (or raising the pH of) an acid soil.

The amount of lime to be applied to soil to achieve a particular change in pH value depends on a number of factors, the most important being the texture of the soil itself. The graph shows the amount of ground limestone to add to different types of soil to raise the pH from various starting values to 6.5, the level most generally useful for garden plants. If you add the lime as quicklime or slaked lime, you should add half of these amounts.

Although I have talked so far of lime solely in respect of its effect in raising pH, it should not be forgotten that it also adds to the soil the

Raising the pH of your garden soil
It is much easier to raise than to lower the pH of soil. By adding calcium carbonate in the form of ground limestone, a considerable change can be effected, although it may take several years with a very acid soil. The amount of limestone that you will need to add to raise the pH from a given value to 6.5, the level suitable for the majority of garden plants, varies with different types of soil but may conveniently be read from the graph.

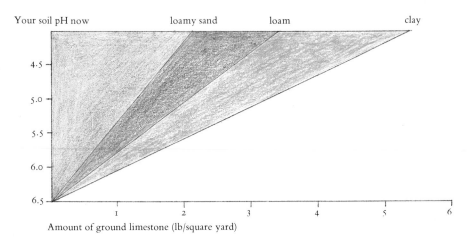

element calcium, one of the essential plant nutrients. If, by chance, plants show symptoms of calcium deficiency, but you do not want to raise the pH, the problem may be alleviated by adding calcium in the form of calcium sulphate (gypsum), which will not have the same elevating effect on the pH. If, moreover, plants show signs of manganese deficiency, as they often do on alkaline soils, then magnesian limestone should be used as a source of both elements.

I have referred to the difficulty of lowering the pH of an already alkaline soil; while logic suggests that, if adding an alkaline chemical will neutralise an acid soil, so adding an acid should neutralise an alkaline one. The main problems arise because appropriate mineral acids, such as sulphuric, are extremely dangerous to handle, and because the use of safer chemicals such as sulphur, that are naturally converted to acid in the soil, is rarely very effective. Although peat is, of course, acidic, very large quantities indeed must be added to have an appreciable effect on soil acidity and, of course, there will be considerable changes in soil structure too in consequence. On balance, I feel that the answers to an alkaline site are to 'go with the soil' and grow plants whose natural ecological niche this is, to use sequestered iron to compensate for any problems in uptake of this element and, if you are quite besotted with acid loving species, to construct a small, separate peat garden for a selection of dwarf forms.

Having nature on your side – how to plan for efficient crop rotation

In discussing pH, I have already touched upon the much larger subject of plant nutrition, for although plants manufacture organic chemicals within their green tissues by photosynthesis (see p. 81), they obtain virtually all their mineral nutrients from the soil. It is on these soil nutrients that the highly unnatural system of plant life known as gardening imposes such a drain. In a natural habitat, a nutritional balance is maintained through the life and death of plants during the ecological succession towards the climax vegetation community, described in Chapter 5. In gardens, we grow alien plants in highly unnatural ways, and most mineral soil nutrients will need artificial replenishment. Irrespective of the form in which they exist in the soil or in which they are applied as fertilisers, the form in which plants actually take up and use nutrients is chemically very simple. The essential elements may be divided into major nutrients and minor nutrients, and their roles in plant growth are summarised in Table 3.

The fertilisers in which these elements may be added to soil can take several forms and the distinctions between organic and inorganic, quick acting and slow release, straight and compound, and liquid and solid fertilisers have important practical implications. Strangely, the distinction between organic and inorganic fertilisers, which seems to hold such a powerful emotive significance for some gardeners, is biologically the least important of the four categories. An organic fertiliser is derived from some once living organism; its name very often defines this – dried blood, fish meal, bone meal, or hoof and horn for example. An inorganic fertiliser is derived from somewhere else, but this need not be a chemical factory, much less a sinister one; it could equally be a hole in the ground. Rock phosphate, basic slag, ammonium sulphate, superphosphate and sodium nitrate are all inorganic fertilisers, and gardeners who baulk at the use of these substances often do so in ignorance of the fact that nitrogen to a plant is nitrogen and, by the time it is absorbed into the tissues, its origin is immaterial. It is also sometimes claimed that organic fertilisers contain quantities of minor or trace elements that inorganic materials

Table 3	Principal nutritional elements in soil and their roles for plant growth	
Nitrogen	A major constituent of protein and essential for many aspects of plant development. Especially important for leafy, as opposed to flower and fruit growth
Phosphorus	A constituent of many proteins, fats and carbohydrates. Especially important in the ripening of fruits, the maturation and germination of seeds and root growth
Potassium	Uncertain, but essential for good flower and fruit formation and probably involved with aspects of photosynthesis and with control of water loss
Calcium	An essential constituent of plant cell walls and in the correct functioning of growing points and other young tissues
Magnesium	A constituent of chlorophyll and essential for the correct functioning of photosynthesis
Iron	Has several functions, but most importantly with the formation of chlorophyll and hence with photosynthesis
Manganese	Uncertain, but in part concerned with the formation of chlorophyll
Boron	At least in part, concerned with rendering calcium available to plant tissues.
Copper	Uncertain but apparently concerned with the correct functioning of some enzymes
Molybdenum	Concerned with rendering nitrogen available to plants from nitrogen-containing compounds in soil, and with root nodule activity on legumes

lack. The counter to this argument is that most soils already contain ample natural reserves of minor nutrients which are not exhausted by cultivation at anything like the rate at which major nutrients are. Moreover, it is often claimed that the minor or trace nutrients present in organic fertilisers result in improved flavour of vegetables and other edible crops. There is scant scientific evidence, however, that such substances have any effect at all on flavour; and the belief that, if these widely disparate chemicals do influence it in this manner they should invariably do so for the better, has always seemed to me quite extraordinary.

The distinction between a straight and a compound fertiliser is merely one of formulation. There are relative few uses for straight fertilisers, organic or inorganic, on their own, because most of them contain too limited a range of nutrients to satisfy all of the plants' needs. They are therefore artificially blended together to produce compound mixtures for specific purposes and specific types of plant.

A slow-release fertiliser is one that, by virtue of its chemical make-up, breaks down in the soil into the simpler form that plants can use over a long period of time – several weeks, if not months. A quick-acting product is almost instantly available to the plants, either because it is already chemically simpler or because it breaks down very rapidly. The practical applications are obvious; if you wish to have a supply of a particular nutrient released gradually to satisfy a plant's needs throughout a season, a slow-release formulation is the answer. If you want more or less instant results (perhaps on very rapidly growing and maturing vegetables), or if a plant has a deficiency that requires immediate correction, a quick-acting product will do this for you. Allied to this is the difference between solid and liquid fertilisers: a liquid will give more rapid results (just as a soluble pain-killer will more quickly cure your headache), as it is already in a form that the system can absorb. Many organic fertilisers, such as bone meal or hoof and horn, are of slow-release action and may, therefore, be valuable on this account if there are no inorganic products with comparable effect and nutrient content.

The fastest results of all are often obtained from a liquid fertiliser applied as a foliar feed, for although most nutrient uptake is through the roots from the soil, plants do possess the ability for limited uptake through their leaves – although, as explained in Chapter 4, the natural ecological applications for this are limited.

Crop rotation

Rotation of cropping in the vegetable plot makes sound sense for several reasons, relating to plant nutrition, pathology and cultivation, as outlined in the text. Nonetheless it must not be expected that the benefits will be as great in a garden as they are in commercial horticulture where the plots can be spaced much further apart. The three course rotation outlined here will be appropriate for a 'self-sufficient' vegetable garden and could of course be modified or adapted to individual cropping requirements. Few gardens are large enough to accommodate a rotation of more than three years.

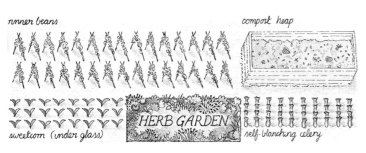

runner beans   compost heap

HERB GARDEN

sweetcorn (under glass)   self-blanching celery

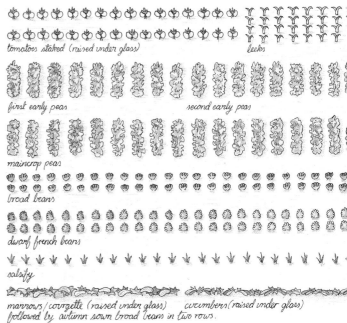

tomatoes staked (raised under glass)   leeks

first early peas   second early peas

maincrop peas

broad beans

dwarf french beans

salsify

marrows/courgette (raised under glass)   cucumbers (raised under glass)
followed by autumn sown broad beans in two rows.

On p. 71, I outlined the reasoning behind crop rotation in terms of minimising the drain on nutrients in the soil. In most vegetable gardens, a three course rotation is followed, the area being divided into three equally-sized plots. On each plot, a different range of vegetables is grown, the siting of each group being changed each year so that the same types of vegetables are only grown on the same soil every three years. Vegetables can be divided into several groups with different needs and, before planning a rotation scheme it is important to appreciate these (amplified with details of some of their natural environments in Chapter 6). Group One comprises those root crops that do not grow well if fresh manure has recently been applied to the land, presumably because they are unable to utilise certain nutrients until they have been reduced to chemically simpler form. Carrots, parsnips, beetroot, Jerusalem artichokes and potatoes fall into this category (though swedes and turnips present less of a problem and are usually considered in Group Two with their close relatives, the brassicas). So it is usual not to apply manure immediately before planting the Group One root crops.

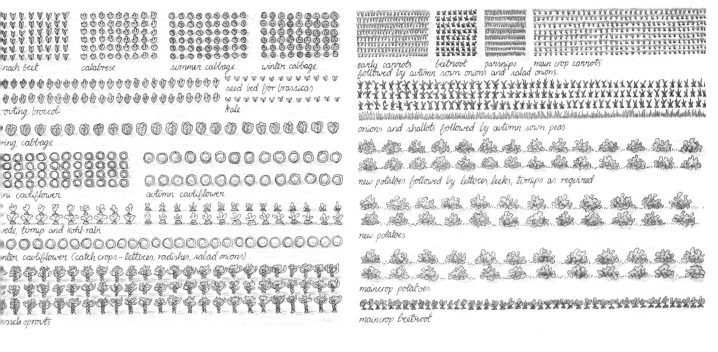

nach beet   calabrese   summer cabbage   winter cabbage

seed bed for brassicas

outing broccoli   kale

ring cabbage

ni cauliflower   autumn cauliflower

ede, turnip and kohl-rabi

nter cauliflower (catch crops - lettuces, radishes, salad onions)

ssels sprouts

early carrots   beetroot   parsnips   main crop carrots
followed by autumn sown onions and salad onions.

onions and shallots followed by autumn sown peas

new potatoes followed by lettuces, leeks, turnips as required

new potatoes

maincrop potatoes

maincrop beetroot

The cabbages, Brussels sprouts, cauliflowers and other brassicas in Group Two generally benefit from a slightly more alkaline soil than do many other vegetables, and one which will inhibit the likelihood of clubroot disease building up in the soil (p. 190). If soil tests indicate that lime is required on your land, therefore, apply it to the plot that is about to carry the brassicas. Brassicas all have a particularly high demand for nitrogen fertiliser; remember that nitrogen is especially important for leafy growth, and most brassicas are very leafy plants.

The crops that comprise Group Three are the legumes; the pea and bean family that is active in actually adding nitrogen to the soil; ideally, therefore, they should precede brassicas in the plan, and, of course, require very little or no additional nitrogen fertiliser themselves.

With these particular groups in mind, a possible three-year rotation is outlined in the diagram. There are clearly other possibilities; indeed, individual families' needs and preferences must also be taken into account. But, so far as it is practicable, the requirements of these three plant groupings should be allowed for.

Table 4
Nutrient content and uses
of the commonest
garden fertilisers

Organic fertilisers

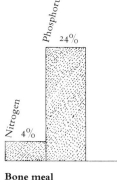

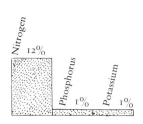

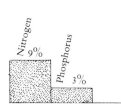

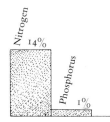

Bone meal
Slow release phosphorus
source for aiding root
growth; contains lime and
therefore unsuitable for lime-
hating plants.

Dried blood
The best quick acting organic
nitrogen source for
promoting rapid leafy
growth.

Fish meal
Fairly quick acting nitrogen
and phosphorus source, often
used in combination with
bone meal and dried blood
as 'Blood, fish and bone'.

Hoof and horn
Slow release nitrogen source
for use in the garden and an
integral component of the
fertiliser mixture used in
John Innes soil-based
composts.

Artificial fertilisers**

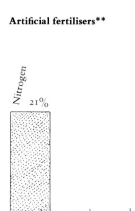

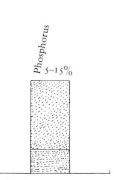

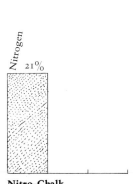

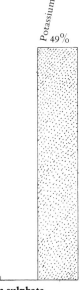

Ammonium sulphate
The most readily available
form of nitrogenous
fertiliser, but removes calcium
from the soil, thus making it
more acidic (see Nitro-
Chalk).

Basic Slag
No longer very easily
obtained but a good source
of phosphorus, especially for
acid soils as it contains lime.
Encourages clover growth on
lawns.

Nitro-Chalk
Contains ammonium nitrate
as a nitrogen source but with
the valuable addition of lime
and especially useful therefore on acid soils.

Potassium chloride
Although containing more
potassium than the sulphate,
it is generally less valuable
because the high chloride
content can damage young
plants.

Potassium sulphate
The best all-round source of
potassium for promoting
flower and fruit development.

*Percentage figures are intended merely to indicate the relative merits of different substances and take no account of the way that the three elements are combined with other chemicals or of the fact that some substances contain a proportion of the elements in forms that are unavailable to plants.

** These, and other, artificial sources of plant nutrient are blended in varying ways to produce compound fertilisers for particular gardening purposes. The best known is Growmore, a fairly quick acting granular fertiliser containing 7% each of nitrogen, phosphorus and potassium, but it is beyond the scope of the present book to describe the many other different mixtures now available.

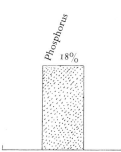

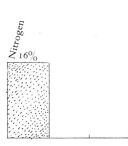

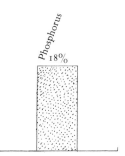

Rock Phosphate
Somewhat similar in composition and uses to bone meal, but contains a useful proportion of magnesium also.

Sodium nitrate
Very fast acting and especially useful on sodium requiring plants such as beetroot and celery. Generally superseded for most garden purposes.

Superphosphate
Similar in use to bone meal and other forms of phosphorus in aiding root establishment and growth, but possibly less beneficial on acid soils.

Chapter Four

Plants and the gardener

Until relatively recently, living organisms were divided for the purposes of school biology syllabuses and everyday usage into two Kingdoms, the Plant and the Animal. This sub-division enabled most people to classify most readily visible organisms in one way or the other; but there were always grey areas when smaller and sub-microscopic groups were considered. Fungi, algae, slime moulds, bacteria, protozoans and, above all, viruses, were among those that created difficulties, and it was more a matter of tradition than scientific reasoning that prompted botanists to embrace one group, and zoologists another. It is now recognised that several taxonomic hierarchies at the level of the old Kingdoms are needed and a typical modern scheme is outlined in Table 5. Representatives of all of these groups may be found playing their role in garden ecology, although you may need to search fairly diligently for some of them. Here, I shall concentrate on 'higher' plants in the modern sense, and shall consider briefly the more important members of what are sometimes called the 'lower' plants' on p. 103. Animal life in the garden is dealt with in Chapter 7.

It is appropriate first to mention two important sub-divisions among higher plants. The first is into conventional flowering plants (Angiosperms) on the one hand and conifers and their allies (Gymnosperms) on the other. The second sub-division is of flowering plants into those that produce only one seed leaf (the Monocotyledons) and those that produce two seed leaves (the Dicotyledons). The former tend to be characterised by narrow, strap-like leaves and include such important families as the grasses, lilies and irises. Structurally, there are some important differences between Monocotyledons and Dicotyledons; but general ecological principles are applicable to both.

Plant structures

A plant has well defined structural and functional parts; the roots, stems, leaves and flowers. While each of these parts is fairly easily recognisable in most plants, a brief examination of any garden will reveal that there is wide variation in the types of each. The bloom on a paeony is very different from the floral spike on a *Kniphofia*, but they are both quite evidently flowers. Even the curd of broccoli and the edible head of a globe artichoke are recognizably floral, but many gardeners would baulk at identifying the true nature of a poinsettia 'bloom' or at even *finding* the flowers of duckweed. Leaves and stems are, in most cases, distinct and fairly self-evident; but cacti, *Ruscus*, the brooms and some

Arum maculatum
*The flowers of the family
Araceae represent one of
many complex adaptations
to encourage and enable
pollination by insects.*

79

conifers might make you think. Roots vary less than other plant parts; but some underground plant structures are modified stems, and some roots can be found many metres up in the air!

The roots, stems, leaves and flowers are a plant's organs. They are composed of separate tissues in regions, each comprising many cells and having a clearly defined function. Few individual plant cells are readily discernible to the naked eye; although they vary greatly in size, the typical cell is an irregularly shaped bag of watery fluid approximately one tenth of a millimetre in diameter, confined by a boundary or wall. Within the fluid, called cytoplasm, float a nucleus, which has overall control of the cell's functions, and various other microscopic bodies responsible for particular vital processes. The boundary wall may be more or less strengthened and rigid, depending upon its particular purpose; the cells of a petal, for instance, clearly do not require the physical strength of those that make up wood. For a plant to grow, it must increase in size

Table 5   The outline of a recently proposed Five-Kingdom classification of living organisms

Kingdom	The more familiar of the major constituent groups
I Monerans	Blue-Green Algae Bacteria
II Protists	Various microscopic, unicellular or colonial organisms including Golden Algae, Amoebae, Ciliates, Euglenoids
III Fungi	All Fungi, including Mushrooms, Toadstools, Moulds, Mildews and Rusts
IV Plants	Red, Brown and Green Algae Stoneworts Mosses and Liverworts Vascular Plants (including Ferns, Horsetails, Conifers and Flowering Plants (Mono- and Di-cotyledons))
V Animals	Sponges Jellyfish Eelworms Earthworms Arthropods (including Insects) Chordates (including Birds, Reptiles, Fish, Amphibians and Mammals)

and this is achieved, not by individual cells becoming larger, but by new cells forming as a result of the division of existing ones. This new growth occurs in particular regions called meristems or growing points and the position of a plant's meristems has some important ecological consequences. For instance, the reason that you cannot produce a lawn from delphiniums is that the meristem is at the tip of the shoots; mow a delphinium and you remove its growing point. A grass, in contrast, grows from the bottom upwards; it has its meristem at the base and, therefore, being mown or browsed by animals will not affect its ability to produce new cells.

There are three particular plant functions that it is important to consider in more detail: those that govern their ability to manufacture their own nutrients, their ability to use these nutrients for growth, and their ability to reproduce.

Already in the book, I have referred to photosynthesis, which can loosely be translated as a food-manufacturing process that occurs in the presence of light. Not only is this phenomenon the manner in which green plants manufacture fairly complex organic nutrients from much simpler chemicals, it is also far and away the most important means by which solar energy is trapped on earth and used to maintain life of all kinds, plant and animal. Although the process itself is quite complex and involves several separate chemical stages, some of which, indeed, take place in the dark, it is quite sufficient for present purposes to appreciate the essential properties needed by a plant before photosynthesis can occur. The structural requirement is for cells containing bodies known as chloroplasts. A chloroplast is a microscopic structure having something of the form of a pile of loose sheets of paper. On each 'sheet' is a chemical called chlorophyll and it is the existence of countless molecules of this one chemical that colours our planet green. Chlorophyll itself is only formed in the presence of light (hence a plant kept in the dark loses its green colour) and its production also requires certain elements from the soil – in particular, iron. Thus, plants that are iron deficient will be pale in colour because they cannot produce chlorophyll satisfactorily. Variegated plants, so beloved of gardeners, are usually mutants having some cells deficient in chlorophyll; hence they generally grow less vigorously and, in the wild state, rarely thrive in competition with their normal relatives.

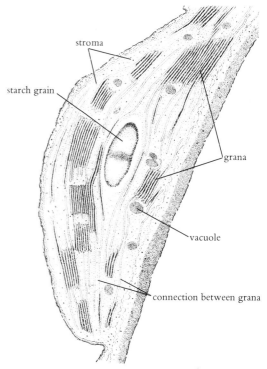

stroma

starch grain

grana

vacuole

connection between grana

Chloroplast structure
The chloroplast is the sub-cellular structure within which the photosynthesis pigment chlorophyll enables a plant to use solar energy to convert simple inorganic chemicals into more complex nutrients. The chlorophyll itself is present on the grana which are minute bodies arranged like the leaves of a book. Such detailed structure can only be revealed with an electron microscope which provided the image on which this drawing is based.

Basic plant function

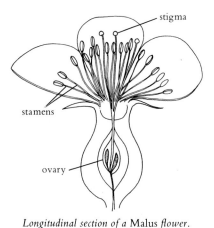

Longitudinal section of a Malus *flower.*

Flowers

In photosynthesis, chlorophyll in the presence of light causes the carbon dioxide, that infuses plant tissues from the air, to combine with water in the cells to produce more complex chemical compounds like starch and sugars. These nutrients are transported around the plant, from the green leaves to other organs, in the specialised cells of the conducting tissues. Another product of photosynthesis is oxygen and although that given off by normal land plants is of little direct consequence in a garden, that produced by submerged aquatic plants can actually be seen as minute bubbles on their leaves and is vitally important if other organisms are to thrive in a garden pool.

The nutrients produced in photosynthesis can be used by the plant in one of three principal ways. First, they may be used, after further chemical change, to produce the raw materials of new cells and so permit the plant to increase in size. Second, they may be stored for future use in a relatively insoluble form; the food stores in tubers, swollen roots, fruits and other plant structures that we make so much use of are really intended to benefit, not us, but either the plant itself (after a period of dormancy) or its offspring if the food store happens to be a seed. Third, the nutrients may be used to provide energy for all these other processes; they may be respired. Respiration is the opposite of photosynthesis; it involves the combination of oxygen with the sugar or other nutrient resulting in the liberation of carbon dioxide, water and the solar energy that has been chemically trapped. Unlike photosynthesis, it takes place in all cells of the plant and is not dependent on light for its operation.

Flowers are plants' reproductive organs and, naturally, all must therefore possess them to enable the species to survive. A few garden plants never produce flowers because they are derived from naturally occurring, mutant, flowerless freaks that, in nature would have died out, but have been artificially propagated and perpetuated in various ways. Many garden flowers, moreover, are unable to produce seed because artificial selection and breeding has resulted in the over-development of some floral parts (those we find most attractive) at the expense of those of more functional significance to the plant itself. The evolutionary origin of the flower was described by Charles Darwin as 'an abominable mystery', and the story is still very imperfectly known. It is generally believed, however, that petals are modified leaves, that the entire flower represents a modified leafy shoot. The Magnolia family

Longitudinal sections of flowers

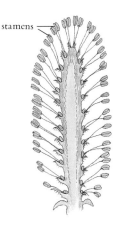

stamens

stigma

ovary

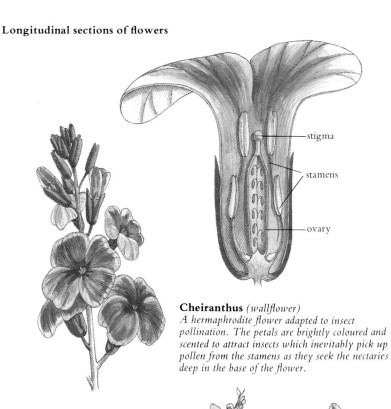

stigma

stamens

ovary

Cheiranthus *(wallflower)*
A hermaphrodite flower adapted to insect pollination. The petals are brightly coloured and scented to attract insects which inevitably pick up pollen from the stamens as they seek the nectaries deep in the base of the flower.

Salix *(willow)*
A wind-pollinated tree with separate male and female flowers. Both stamens and stigmas are exposed, respectively to discharge and receive pollen which is produced in large quantities. The flowers of wind-pollinated plants are usually sombre in colour and lack scent.

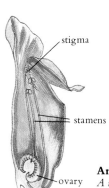

stigma

stamens

ovary

Antirrhinum *(snapdragon)*
A more complex adaptation to insect pollination in a hermaphrodite flower. Only large and powerful insects like bumble bees are able to force apart the tensioned petals and so gain access to the interior of the flower.

83

(whose flowers are closely similar in structure to leafy shoots) is considered the most similar to ancestral types among surviving forms.

Most of the enormous variation seen today among the flowers of wild plants represents evolutionary adaptations to different methods of pollination. Some flowers are fertilised by their own pollen, others by pollen from another flower on the same plant, and many by cross-pollination from a flower on another plant. In some self-pollinated flowers, the transfer of pollen from stamens to style within the same flower is a purely mechanical process. In other instances, and certainly in all those where pollen is carried from one flower to another, some method of transfer is necessary. Wind is an obvious and common agent; water, for aquatic plants, is another; but most flowering plants depend on some form of animal life, most importantly insects. Few wind-pollinated flowers (many are commonly called catkins) are very colourful, and their appeal for garden ornamentation lies more in their delicate shape and form. Insect-pollinated flowers, on the other hand, are usually much more brightly coloured, and often scented too, simply to attract particular species of insect. Two extreme examples are illustrated of garden flowers whose structure is adapted to facilitate pollination by insects. But pollination mechanisms can vary enormously even within closely related flowers: within the phlox family, for instance, are species pollinated by bees, various types of fly, beetles, butterflies, both day- and night-flying moths, birds and bats, as well as many that pollinate themselves.

While good and successful pollination, resulting in seed set and fruit production, is seldom of significance to the grower of ornamental garden plants, it is obviously very important with fruiting forms. And knowledge of pollination ecology can help ensure a good crop; insect-pollinated plants such as raspberries and blackberries, for instance, may produce few fruits on a windy site where the insects will be blown away; provision of a little shelter may make a great deal of difference. (Indeed, one of the reasons why plants that grow naturally in very windy places – mountain tops or open moorland, for example, are low growing is partly to ensure that their flowers are down where pollinating insects can survive.)

Conversely, wind-pollinated plants are often best planted in small groups, to ensure that plenty of pollen is available, and placed where the wind blows gently through the group of individuals.

On many garden plants, the flowers are present individually (paeonies, cucumbers and daffodils, for instance); on others they are loosely grouped together (tomatoes, apples and aquilegias), whereas in many (laburnum, kniphofia, onion), they aggregate to form a head of quite distinct form. In the most extreme type of aggregation, seen in the daisy family, the masses of tiny flowers themselves vary in shape, form and function to produce a composite head, superficially very similar to an individual bloom. All of these floral types are adaptations to particular pollen transfer systems.

Seeds

The end result of flower growth and pollination, at least in plants unaltered by cultivation, is the formation of seeds. The quantities of seed produced by an individual plant vary greatly, as do the forms and sizes of the fruit that contains them. At the extremes are the fruit of the coco-de-mer palm of the Seychelles, with a single seed weighing up to 20 kg, and those of some tropical orchids containing masses of individual seeds, each of less than one millionth of a gram. As with pollen, many and varied are the methods by which seeds are dispersed from the parent plant; wind, water and animals of all kinds may be involved, but the fact that seed becomes dispersed at all is of great ecological importance. Seeds landing and germinating close to the parent will be in competition with it and with each other for light, nutrient and water. By travelling some distance away, they not only avoid this difficulty but also enable the species as a whole to colonise new territory. The extreme example of dispersal is again a species of palm; the coconut, which has spread right around the warmer parts of the world through the simple expedient of having seeds that can float on, and not be harmed by, sea water. Seeds are also the means for the survival of species during adverse conditions when the parent plant has died. This survival of the ungerminated seed may simply be the year to year process of annuals, the span of several years familiar with so many garden weeds (see p. 117), or the several hundred, and, possibly thousand years claimed for a few species.

A distinction must be made with regard to one of the groups of higher plants commonly grown in gardens, the conifers and their allies.

While these plants produce seeds, they are quite primitive in the evolutionary sense and bear their seeds, not in fruits, but in cones. The precursor of the mature cone, in which the pollen is borne, is superficially rather like a conventional flower, but structurally quite distinct. In some ways, these plants are intermediate between the spore-bearing ferns and related groups and the typical flowering plants. Nonetheless, many of the general ecological principles relating to seed formation are applicable to conifers also.

Stems

A plant's stem may be thought of as its skeleton, or even its body, holding the leaves and flowers in their correct functional positions, and providing the channel for the transportation to and from these organs of nutrients and water. Although it may contribute in a minor way to photosynthesis, it is primarily a mechanical feature and one, moreover, that endows plants with the various shapes and modes of growth (known as ecological life forms) that I consider later.

Despite their enormous variation in form, all stems possess the same basic structure, with one or more strands of conducting tissue through which water and nutrients move, a surrounding zone of 'packing' tissue with provision for the radial movement of water and nutrients, and an outer 'skin' called the epidermis. One aspect of a plant's stems and stem tissues above all, however, confers ecological status; their ability to produce an extremely durable chemical called lignin and to develop the feature known as secondary thickening. In everyday terms, this really means the ability of the plant to form woody tissue; compare the stem of a *Calendula* with a one-year-old rose shoot and the trunk of an oak tree and you will see the result of an increasing content of lignin in the tissues. (Monocotyledonous plants do not produce wood in the conventional sense, but some, such as palms and bamboos, do have analogous ways of strengthening their stems.)

There are two important ecological consequences of the development of lignified tissue; first, because such tissues are resistant to dessication and decay (especially when combined with a comparably thickened epidermis in the form of a layer of bark), they provide a plant with the means of surviving above ground during periods of adverse conditions, such as over a cold winter or during a drought. Thus, while many perennial plants must perforce die down each winter to a bulb, tuber or other underground body, a perennial plant with woody tissues can remain above ground. However, because its dormant tissues are not buffered by the soil from the extremes of temperature, an exotic plant grown in a harsher environment than its native climate may need additional protection in the form of a screen of straw or other material.

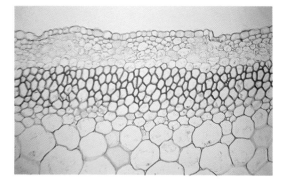

A microscope section cut transversely through a stem reveals the wide range in shapes, sizes and wall thicknesses of cells. Each type has its own characteristic structure and function.

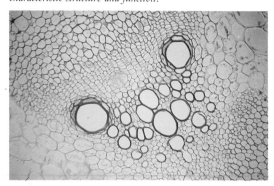

Well protected towards the centre of the stem are the conducting tissues that carry nutrients and water to all parts of the plant. The wide water-conducting cells are especially evident.

The Coast Redwood,
Sequoia sempervirens,
seen here in its native
California, reaches a greater
height and weight than
almost any other tree; a feat
achieved through the
durability and structural
strength of the chemical
lignin.

The ability to survive its period of dormancy above ground confers a great ecological advantage in that a plant doesn't have to start its growth 'from scratch' each season. It can capitalise on the results of the previous and earlier years and, of course, grow much larger in consequence. All of the largest forms of plant life have this woody stem structure that can result, in some conifers and eucalypts, in heights of over 100 m, a mass of over 2,000 tonnes and, in *Pinus longaeva*, the bristlecone pine, in an age exceeding 5,000 years. Few small plants possess woody tissues and those that do tend to occur in harsh climates and to be very slow-growing. Conversely, the really large plants with non-woody tissues usually occur in moist, tropical climates where a rest period is not essential, and where ultimate size tends to be limited by physical instability: such plants invariably require the support of other vegetation for their weak stems.

Leaves

Some school botany textbooks refer to the leaves as the 'powerhouse' or 'factory' of the plant – fanciful, perhaps, but illustrative of their primary function of trapping the sun's energy and using it in photosynthesis. Leaves vary enormously in shape, size, disposition on the stem and, if the green chlorophyll is partially absent or obscured by other pigments, in colour. The ecology of the leaf is essentially one of a structural compromise between presenting a surface that maximises the capture of solar energy, with one that minimises water loss. For the second important leaf function is as a medium for gaseous exchange; all leaves possess pores (or stomata) through which air – containing the essentials of carbon dioxide for photosynthesis and oxygen for respiration – passes inwards, and water taken in from the soil passes outwards as water vapour. The balance between adequate and excess water is considered later.

Despite the vast variation in form between leaves, they all, like stems, possess strands of conducting tissue, 'packing' tissue within which the photosynthesis occurs, and an outer epidermis. The total surface area of the leaf (and, of course, the total number of leaves per plant) tends to reflect the energy-trapping capability; rather like a solar panel on a house roof, a large, flat surface facing the sun is the most efficient. The shape of the leaf, on the other hand, together with the thickness of the outer cell-walls of the epidermis, tends to be an adaptation to limit water loss. Size and shape can also have other functions: the needles of conifers in northern regions may be arranged to facilitate the sliding off of snow; the holes in the leaves of some plants are to limit wind damage; and the modified leaves of cacti, reduced to spines, deter animals from browsing.

The distinction between deciduous and evergreen plants (or, more literally, between those that shed all their leaves simultaneously each year and those that retain them for longer periods and shed them piecemeal) has little basic botanical significance. There are genera like *Magnolia* that contain both deciduous and evergreen species, although in most groups one system predominates. Almost all conifers, for example, are evergreen; but the larches are notable exceptions. Ecologically, the evergreen habit has some important consequences; in a mixed vegetation community, an evergreen species will cast shade all year round. And evergreens are more prone to the effects of winter wind or snow damage, because they present a larger surface area; but conversely they have a 'head start' with growth in the spring. Why the deciduous condition should have evolved from the evergreen is a matter of controversy, but it may have been in a complicated response to moderate (but not severe) seasonal cold, high summer humidity and other factors.

There are numerous ways in which leaves have become structurally adapted to effect the compromise between heat-gathering and water loss, and a few instances will illustrate some of the general principles. In moist conditions with low light intensity, leaves tend to have a high ratio of surface area to volume for maximum energy trapping and to have a thin cuticle or covering to the epidermis; the deeply dissected and fragile leaves of many of our native woodland plants illustrate this well. In dry climates with high light intensity, however, the need to limit water loss takes priority over the need to trap solar energy and the leaves of many desert plants are correspondingly small, relatively thick, and possess a thick, water-retentive cuticle. So too, are those of many alpines, for the brilliant sunlight and drying winds of the mountain top can result in much the same problem for a plant as conditions in a hot desert. Leaves may be curled downwards (to the extent, in the extreme form of conifer leaves, of being almost hypodermic-like) in order to limit water loss from the stomata that lie on the lower surface. When confronted with any plant new to him, therefore, a gardener should give more than a passing glance to its leaves: they may well tell him a great deal about the environment in which that plant grows naturally, and indicate something of the conditions that it will need in cultivation.

Roots

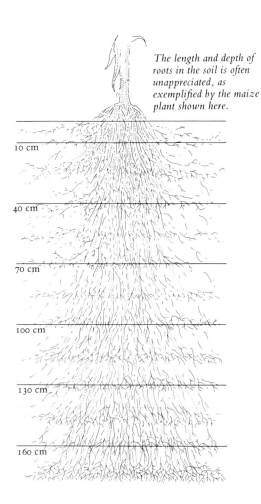

The length and depth of roots in the soil is often unappreciated, as exemplified by the maize plant shown here.

10 cm

40 cm

70 cm

100 cm

130 cm

160 cm

Whether they grow in soil, on the bark of forest trees, or float on water, almost all flowering plants have roots. Principally, they provide physical stability and a means by which water and mineral nutrients can be taken up from the soil. The internal appearance of the individual root is not, at first sight, recognisably very different from that of the stem, having central conducting tissues, surrounding 'packing' tissue and an outer epidermis. It differs structurally, however, in that it lacks any cells containing the green photosynthesis pigment, chlorophyll, and behaviourally in that it grows away from light and towards moisture. Although the process varies between Monocotyledons and Dicotyledons, there are also mechanisms by which older roots, like stems, can be reinforced, although the physical strength possessed by the stem is not, of course, required below ground.

The extent of a plant's root system is not usually appreciated by gardeners, although it is evident when a tree is dug up that there is growth in the soil not very different in extent from that above ground. Nonetheless, it comes as a surprise to many to appreciate that beneath a mature vegetable plot, there may well be over 20 km of roots per square metre.

Although the basic distinction between a long, vertically penetrating tap root, such as that of a carrot, and the many branched fibrous adventitious roots of a grass will be familiar, the extent and degree of branching of a plant's root system are also influenced by the soil structure, texture and depth. Although few gardens possess insufficient depth of soil to prevent the development of the relatively short tap roots of herbaceous plants, some may well be unsuited to the growth of deeply rooting trees; the nature of a plant's root system is one criterion for selecting species appropriate to your garden.

A subsidiary function of the root system of some plants is as a food reserve. Thus, many non-woody perennials such as dahlias or celandines develop tubers or other root structures in order to permit above-ground growth to start afresh at the start of the season. (It should, however, be appreciated that many of the familiar overwintering food reserve bodies, such as bulbs and corms are not roots, but modified stems.)

While the visible roots of a plant provide the obvious means of physical anchorage, the vital nutrient- and water-absorbing function of a root system is performed by structures that, individually, are virtually invisible to the naked eye, the root hairs. These tiny extensions from

The dispersal of fruits and seeds
Many and varied are the mechanisms by which plants disperse their fruits and seeds. Those shown here were collected within the space of a few metres of English hedgerow in early autumn. They include adaptations for being blown long distances by the wind (clematis and willowherb), for adhering to animal fur and hair (cleavers and teasel), for being eaten entire by birds, the seeds surviving unharmed in the digestive tract (rose), and for simply being shaken out, pepperpot fashion (campion).

cells of the epidermis will be apparent if a seed is germinated in a moist atmosphere and the young root examined closely. *En masse*, they give the root the appearance of a minute paint brush and it is then that their delicate nature can be appreciated. It is then, too, that it will be understood why transplanting is such a traumatic experience for plants. In such an upheaval, the root hairs are irreparably damaged and only if the operation is performed into warm, damp soil, or at a time when the plant is dormant and thus not requiring immediate water uptake to replenish that being lost through the leaves, will there be opportunity for new root hairs to form before the plant as a whole suffers materially.

Plant reproduction

Although, in a garden, we tend to think of plants as individuals and, indeed, think in terms of a single season at a time, nature takes a very much more long-term view of matters. What is really important is not how well an individual plant grows but how satisfactorily it can contribute to the survival of its species as a whole. It may be of very great concern to us if a treasured maple succumbs to a gale; but if that tree, during its life, has produced a good quantity of seeds, then it will have served its purpose in the natural scheme of things. Plants as individuals are expendable. It would be a large and rather unusual garden, however, in which the continuation of plant life was solely as a result of natural processes; to wait for one of its seedlings to reach maturity as a replacement for the lost maple would take a patient gardener indeed. Nonetheless, there are gardening situations in which natural multiplication and reproduction are important, but before outlining the relative roles of these *vis à vis* artificially manipulated methods, I must first describe briefly the various types of plant reproduction.

The normal, natural method of plant reproduction is through the medium of seeds, bodies containing an embryo plant with a small supply of food to enable it to grow to a size where it can develop leaves and thus begin to manufacture its own nutrients. The embryo and the seed arise as a result of the fertilisation of an ovule by a pollen tube. Some seeds, such as those of many desert plants, are equipped with very durable outer coats to enable them to survive long periods of adverse conditions in a dormant state. When grown in gardens, a certain degree of horticultural trickery may be needed to break this dormancy and induce germination to take place. Other seeds, like those of *Lobelia*,

A swelling at the top of the trunk is so familiar a feature of many ornamental trees that it is often taken for granted as natural; it actually indicates uneven growth rates between rootstock and grafted variety.

designed to germinate rapidly after falling onto the soil surface, have very small nutrient reserves and if a gardener buries them in his compost, the resultant seedlings will die before they reach the surface.

A gardener wishing to engage in the breeding of new plant varieties will want to encourage seed production in his chosen plants; more commonly, however, it is the survival of the individual that is his concern and the removal of dead flower heads and the preventing of seed maturation are important so that the plant's energy may be directed into building up its vegetative form in order to flower well in subsequent seasons. Seed production is an energy-sapping process for a plant, to the extent that many species expire totally at the end of it. This is, of course, true of all annuals; but certain long-lived perennials, such as some agaves, flower and seed only once (after reaching a great age) and then die – a rather unfortunate happening if such a plant occupies a crucial position in your garden at the time.

In the wilder parts of many gardens, the self-seeding of certain annuals is to be welcomed for, in this way, an ever-present population can be built up. In other instances, if the annual happens to fall into the group of plants known as weeds, the process is rather less desirable and I shall discuss seed production and survival among weed species in Chapter 5. Even if you wish to intervene, collect seed from cultivated plants and sow it elsewhere in the garden, it very often pays to sow it promptly after collection for, at that time, many seeds have not yet embarked on the dormant state that may make germination more difficult later.

Many garden plants, even if they produce flowers, never produce seeds, because selective breeding may have resulted in functionless blooms as described earlier. That there are a few wild plants that rarely produce seeds and yet are very successful in other respects (some strains of celandine and couch grass, for example) reminds us, however, that there are alternative methods of reproduction open to plants and upon which we can capitalise. While the sexual reproductive process is essential for new strains to arise through genetic recombination, a species can multiply quite simply through one individual growing large and then shedding parts of itself. This is known as vegetative reproduction. The creeping underground stems known as rhizomes, the arching above-ground stolons and horizontal runners that root wherever they touch down, the stems that run horizontally along the ground and then turn upwards to form off-sets, are all examples of the numerous ways

Vegetative propagation

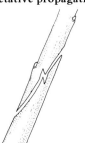 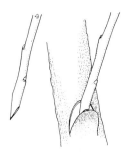   

Whip and tongue graft
Often used on young fruit tree stocks where the stock (below) and scion (above) are of similar thickness.

Side graft
A useful method for rejuvenating a tree by inserting a new scion into an old stock framework.

Cleft graft
Used where the stock is of much greater thickness than the scion. A simple method but leaving an open wound.

Splice graft
Similar to the whip and tongue technique but simpler and relying on secure binding for success.

Crown graft
More sophisticated than cleft grafting and often used to rejuvenate worn out fruit trees.

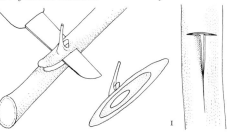 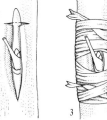 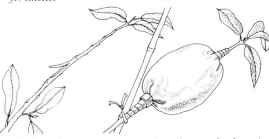

Budding
A bud is a small graft that makes use of one bud only and is thus very economical of plant material. **1.** *A small shallow 'T'-shaped slit*

should be cut in the bark of the stock and gently lifted with the tip of the knife. **2.** *The bark sliver bearing the single bud should be carefully inserted into the*

slit, ensuring that the bud itself faces upwards. **3.** *Binding is as important with budding as with other forms of grafting to hold the tissues firmly and create good adhesion.*

Air layering
A useful if unsightly method of inducing a shoot to produce roots while attached to the plant.

A small cut on the shoots is wedged open, packed around with Sphagnum moss and the whole wrapped in plastic to retain moisture.

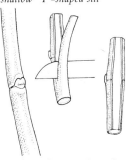 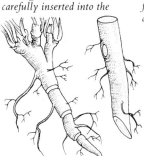   

Bud cuttings or 'eyes'
These are the traditional method of propagating vines but the exposed tissues must be kept moist.

Root cuttings
Unlike other cuttings, these are taken during the winter. The slanting cut indicates the bottom, and the straight cut the top.

Pipings
These are the shoot tips pulled rather than cut from carnations and pinks, and potted up in the usual way.

Leaf bud cutting
Blackberries and camellias especially are most simply propagated in this way; the bud should be half buried.

a. *Nodal cutting*
b. *Heeled cutting*
Some cuttings need a small portion of old wood, the heel, to root successfully.

An open habitat
In this garden, the centre is open – there is a large area of lawn, and the beds and borders adjoining it also receive a high degree of illumination.

Plant life forms

that individual plants can spread further into their immediate environment. Sever the link between the parent plant and the offspring, either by natural decay processes or by the intervention of the spade or the secateurs and you have increased the number of individuals. Not as fast or as far as by seeding, perhaps, but very efficient nonetheless, for there is no need for the new plant to pass through very hazardous dispersal, germination and seedling stages.

All the vegetative methods of reproduction can, of course, be encouraged and used in the garden. There remain, nonetheless, a few purely artificial methods of plant multiplication that must be mentioned; they are very important in gardening, although scarcely relevant in a natural plant community. I have referred to the meristem or growing point of a plant as the region where new cells form. It is, however, important to realise first that the cells produced by the growing point may be very different from the overall tissue in which the growing point is situated, and second that new growing points can themselves arise in tissues. Thus a length of stem, cut from a growing plant and thrust into the ground, may well begin to produce roots and so form a new individual. Hence the theory of taking cuttings, a very valuable operation for the rapid multiplication of plants that produce no seed. For even more rapid progress towards a mature plant, plunge the cutting into an already growing rootstock of a closely related plant and you have the basis of grafting and budding.

It may be self-evident and seem slightly ludicrous to say so, but all plants are different. Clearly, there are differences between species and varieties, and indeed between individuals of the same species. Nonetheless, there are differences that transcend all of these categories and that will be apparent to anyone with no knowledge of plants at all. Show a child a group of plants and he will describe them as big, small, flat, twisted, tufted, perhaps soft and hard, stiff or bendy, prickly or smooth. The more knowledgeable might replace these juvenile terms with such expressions as rosettes, bulbs, trees or shrubs. These are all manifestations of adaptation to particular ecological circumstances. They are governed by the ranges of form and structure displayed by the flowers, stems, leaves and roots, and, expanded through a knowledge of plants' life histories, described and defined in rather more precise terms, they represent examples of what are known as plant life forms.

A closed habitat
Here, in a woodland garden, there is a fairly closed canopy formed by the trees and, as a result, relatively little sunlight reaches ground level.

The system of naming plant life forms most widely used by ecologists today is based on that devised in the early years of this century by the Danish botanist, Raunkiaer. The basis of this scheme is the position on the plant of the buds that are the means of survival during periods of dormancy. The system contains four main groups of perennial plants, each subdivided further, together with annual plants that complete their life cycle from seed to seed within one season, have no perennial buds at all, and are, as individuals, transient members of a community. The groups range from tall trees, with their buds well above ground level, through plants having the buds increasingly closer to the ground, to those whose dormant buds are actually buried in soil, mud or the floor of some aquatic habitat. The names that Raunkiaer used for the different categories are long and unwieldy and I have translated them liberally (Table 6).

It is now pertinent to examine a few examples of widely contrasting geographical regions to see the proportions in which the different life forms exist within their native floras, and why this might be (Table 7). In a typical western European climatic environment, such as that of Britain, the predominant plant life forms are those with survival buds at ground level and there are relatively few species of trees and shrubs. (As will be apparent from the next chapter, the numerical importance of trees and shrubs in our native flora was once much greater, but the total numbers of species of the various groups has not shifted much.) The advantage to a plant of having its buds at ground level is obvious in an environment such as that of modern Britain where grazing animals abound and where anything taller is likely to be browsed off. In more mountainous parts of Britain, such as the Scottish Highlands, there is, however, a shift towards the bushy, flopping, creeping and cushion type of vegetation; these are the fairly low growing plants that dominate the Arctic flora where the deep winter snow affords protection to their buds, exposed just above the soil. In the short summer, the air close to ground level soon warms, and leaf and flower development can take place rapidly. In complete contrast is the tropical rain forest flora of Mauritius, dominated by trees and with a sprinkling of the epiphytic and stem succulent plants like many orchids that scarcely occur outside the tropics. In this warm, moist environment, growing conditions are optimal and, with few climatic difficulties to overcome, vegetation can truly, in the popular term, become luxuriant. The other

Table 6	**Life forms of plants**	Examples among garden plants
1 Survival buds borne on shoots well above ground level		
	a Trees and shrubs (evergreen or deciduous) Large trees $>$ 30m tall Medium trees 8–30m tall Small trees 2–8m tall Shrubs $<$ 2m tall	Cedars, cypresses, beech Holly, birches, maples Juniper, cherry, crab apple Rosemary, potentilla, hydrangea
	b Epiphytes (plants growing on other plants, especially on trees)	Some orchids and bromeliads, air plants
	c Stem succulents (plants with large, fleshy stems)	Cacti, some senecios and euphorbias
2 Survival buds very close to ground level		
	a Bushy plants (aerial shoots die away to leave buds at stem base)	Michaelmas daisy, border phlox, solidago
	b Flopping plants (aerial shoots flop over so that buds are protected close to the ground)	Some clematis, stellarias and labiates
	c Creeping plants (aerial shoots creep and root at ground level)	Many thymes and campanulas, creeping jenny
	d Cushion plants (very compact creeping plants, confined to a cushion form)	Thrift, many saxifrages, some artemisias
3 Survival buds at ground level, all aerial parts die away		
	a Basal bud plants (survival buds arise at ground level)	Raspberry, ceratostigma, gypsophila
	b Partial rosette plants (most leaves are in a basal rosette, but a few form on the aerial stem)	Chamomile, bugle, carnation
	c True rosette plants (all leaves are present in a rosette at the base of the aerial stem)	Some saxifrages, dandelion, daisy

4 Survival buds form below ground level or under water

<table>
<tr><td></td><td>a Earth plants (survival buds present on a rhizome, bulb, corm or tuber)</td><td>Iris, crocus, dahlia</td></tr>
<tr><td> 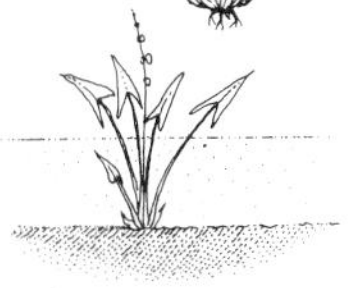</td><td>b Mud plants (survival buds are submerged in aquatic mud but aerial shoots extend above water level)</td><td>Water plantain, reedmace, arrowhead</td></tr>
<tr><td> 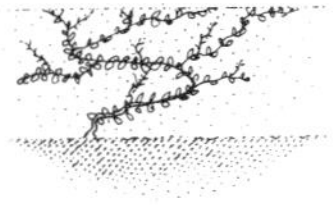</td><td>c Water plants (survival buds are borne below water and aerial shoots do not extend above water level)</td><td>Water lily, Canadian pondweed, hornwort</td></tr>
</table>

5 No survival buds; survival as seeds

    a Annuals   Lobelia, calendula, radish

Table 7 **Percentages of different life forms among the total species of various climatic regions**

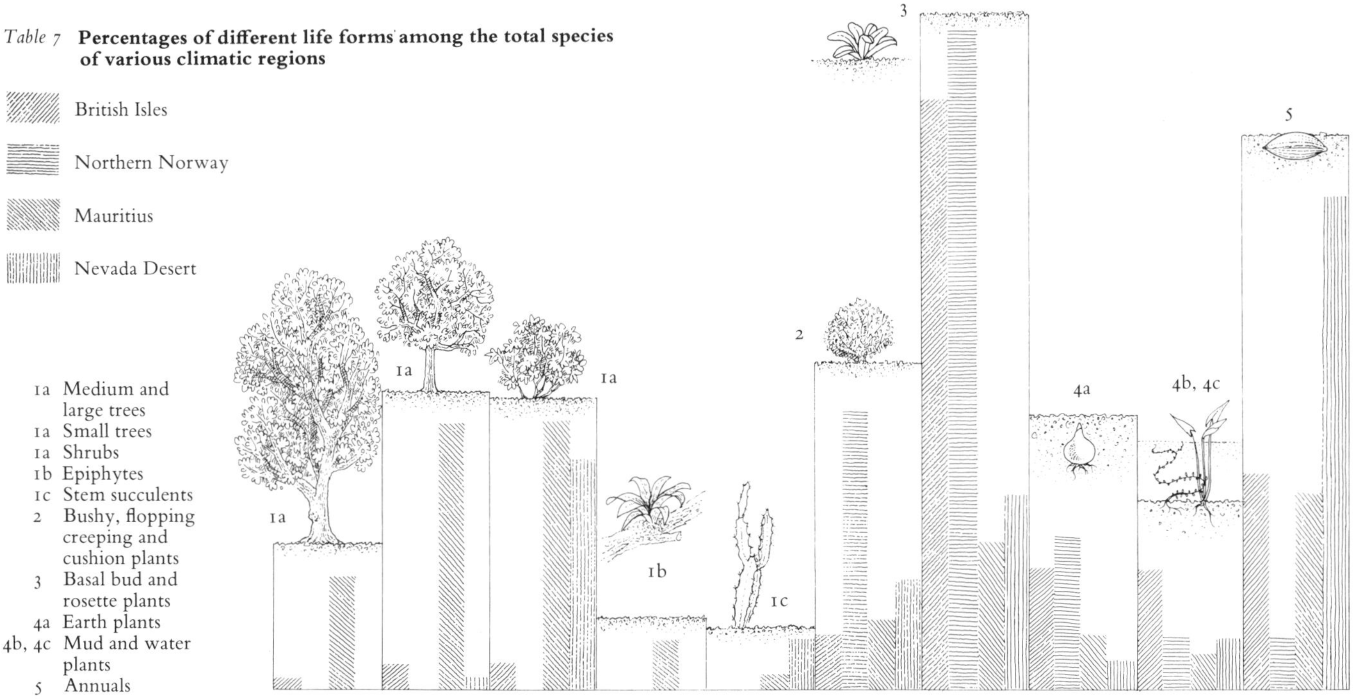

British Isles

Northern Norway

Mauritius

Nevada Desert

1a Medium and large trees
1a Small trees
1a Shrubs
1b Epiphytes
1c Stem succulents
2 Bushy, flopping creeping and cushion plants
3 Basal bud and rosette plants
4a Earth plants
4b, 4c Mud and water plants
5 Annuals

97

The variation among plant morphologies and life forms is enormous. Contrast the climbing shrub form of the passion flower (above) with the small tree habit of the Acer (above right) and the classic rosette form of the Plantago (right). Despite the apparently limitless differences, it is nonetheless possible to categorise all forms of flowering plants into a fairly small number of groupings, based on the position in relation to ground level occupied by the dormant winter buds. This system of life form classification was devised originally by the Danish botanist Raunkiaer and is still used extensively at the present time. Analysis of the component species of the floras of different parts of the world reveals marked differences in the proportions of each life form among the total (see p. 97 for four contrasting examples).

extreme is seen in the Nevada Desert, where very few plants survive above ground for very long; annuals are predominant, capitalising on the short, occasional rains to complete their life cycle from seed in seed in a matter of only a few weeks.

While the pattern of life form distribution clearly reflects major climatic effects on plants in different regions, the composition of individual plant communities may not follow these general patterns: small, local, climatic, soil or other influences will over-ride them. In Chapter I, I introduced the concept of the plant community, and it is now appropriate to look at how such a community is structured and how it can be analysed. Any understanding of garden ecology must be prefaced by an appreciation of the way in which, through their varied life forms, different species make different demands upon their natural environments, utilise different soil horizons, respond to different light levels, react to winter cold, and generally interact with each other to mutual advantage. By considering the structure of a natural plant community, a gardener may appreciate better why certain plant combinations will never succeed in his garden, and why problems arise when certain species are grown in monoculture, away from their natural companions.

There are three ways of looking at the composition of a plant community; first, in terms of the relative numbers of individuals of different species per unit area. This, of course, tells you nothing about the effects that the various species have on each other. To take a garden example, a flower bed might be described as containing three types of plant thus:

Dianthus sp. – Abundant
Lobelia sp. – Frequent
Taraxacum officinale – Very Rare

which gives no impression of the fact that one hundred carnation plants are edged all round with lobelia and that the effect is somewhat diminished by a solitary, large dandelion in the centre!

A second way to approach the matter is by mapping the horizontal distribution of the various plants. This would indicate very graphically the relative roles of the carnations, lobelia and our dandelion, and it is the method generally used by garden designers when seeking to plan herbaceous borders, vegetable plots and shrubberies. In a natural habitat, it also gives a very good impression of communities that are relatively simple in terms of floristic composition, especially if they are mostly

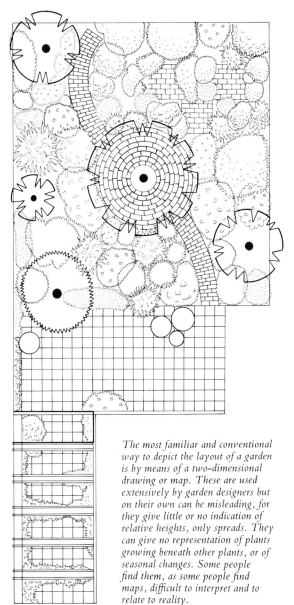

The most familiar and conventional way to depict the layout of a garden is by means of a two-dimensional drawing or map. These are used extensively by garden designers but on their own can be misleading, for they give little or no indication of relative heights, only spreads. They can give no representation of plants growing beneath other plants, or of seasonal changes. Some people find them, as some people find maps, difficult to interpret and to relate to reality.

foxglove   ivy   St John's wort   grasses   bramble   ferns   rowan

     hawthorn   oak   birches   holly   oak   hazel   rosebay willowher

Unlike the garden plan shown on the previous page, this vertical section through an oakwood gives a clear indication of the structure of the community; the small trees, shrubs and herbs would be visible on a map. While this natural assemblage of plants clearly has more tall components than does the garden, a combination of both types of representation gives an immeasurably more telling picture of the make-up of any group of plants.

low-growing. Horizontal mapping falls down, however, in what is usually called a layered or stratified community, such as a woodland, where some plants grow beneath others and where there are many different species of varying heights. Here, the best way to map the plants is by means of a profile diagram. This is a concept new to most gardeners and although it will immediately have relevance for those fortunate enough to own small areas of woodland, the notion of 'thinking vertically' can be applied to other garden habitats too. The two examples illustrated show how plants of different life forms, structures and environmental requirements can interact to create three-dimensional communities in both natural and artificial situations.

Within a classically stratified community such as natural oak woodland, it will be apparent that several fairly distinct layers can be distinguished; the tree layer, the shrub layer, what is usually called the field layer and, finally, the ground layer. It will also be evident that, the shorter the plant, the more is it shaded by the layers above it and, therefore, to thrive, the more shade tolerant it must be. On the other hand, the low-growing plant is better protected from the drying effects of wind and sunshine and, as I mentioned on p. 89, it can afford the luxury of large, fragile leaves in order to capture as much as possible of the available solar energy. The oak tree, with its crown above the remaining plants, is buffered by strong, but pliable branches against wind damage, by fairly small and 'thick-skinned' leaves against drying out, and yet at the same time its very large total leaf area is effective in gathering energy.

The perceptive among you will have noticed that my oak wood is depicted in mid-summer: the bluebells and the dog's-mercury have gone from the field layer. This raises a very important factor when we come to translate ecological principles to gardening: within any plant community, natural or artificial, there are changes within a season. To pursue the oak wood example: clearly, the plants in the field and ground layers that have their active growth early in the season will be in much less deep shade than those flourishing later, after the trees have come into leaf. On p. 47, I demonstrated the seasonal changes in temperature within a garden flower bed, and in other microclimatic factors within a meadow. Comparable changes occur throughout the season in the wood, and the effect of these on the composition of the field layer flora are shown in Diag. p. 102. A plant community is a dynamic blend of different species, each taking turns to fill particular ecological niches.

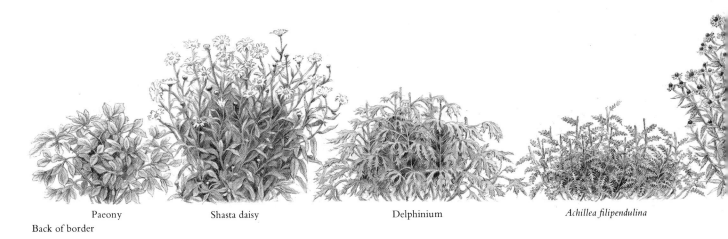

Paeony   Shasta daisy   Delphinium   *Achillea filipendulina*

Back of border

The conventional advice given repeatedly in many popular gardening books for planning herbaceous borders is to place the tall plants at the back, the less tall plants in the centre and the shortest plants at the front. After a few seasons of gardening, most gardeners realise the shortcomings in this advice, as a glance at the vertical section through a herbaceous border shown above will demonstrate.

Leafing and flowering times of some field layer woodland plants
Adaptations to different light requirements mean that some woodland species come into leaf (green) and flower (pink) before the tree leaf canopy is fully developed, whereas others, more shade tolerant, do so later. Gardeners should be aware of this when underplanting.
(Based on an original by E. J. Salisbury)

	January	February	March	April	May	June	July	August	September	October
Dog's mercury										
Lesser celandine										
Bluebell										
Wood anemone										
Pignut										

Helenium   *Ceratostigma willmottianum*   *Sedum spectabile*   *Aster thomsoni nana*   *Alyssum saxatile*   *Nepeta*
                                                                                                                            Front of border

*In this border, some of the plants at the back
are much shorter than those in front of them and
this is simply because not all grow and flower
at the same time. The border is shown in late
summer when the paeonies, for example, which
dominated the border in the spring, are merely
mounds of leaves. The skilled gardener learns
(by the experience of his own mistakes) to allow
for the seasonal changes in his garden when
planning his planting schemes.*

If we now turn to the second profile diagram, through a herbaceous
border at the same time of year as the oak wood, it is clear that no-one
taller than the family cat is likely to be able to stroll through and appreciate much of a vertical distribution of plants at any one point. Because of
this, and because of the requirement for the whole to appear attractive
for as much of the year as possible, the seasonal changes in flowering
patterns of the component plants are of paramount importance. Thus,
the skill of the gardener lies in knowing intimately the times of year that
countless alien and hybridised plants will flower in our climate, in
understanding their relative growth-rates, the details of their shade,
light, heat and drought tolerances, their soil and nutritional preferences,
as well as their aesthetic appeal when blended with others of quite
different colour and shape. Small wonder that the skilled garden
designer is born, not made!

A little thought will indicate another feature of the oak wood and
the herbaceous border. Not only are they frozen in time in the sense of
being depicted at a particular point in the season, but they are also
frozen at a certain stage in their own longer history. In ten years time,
both habitats will reveal different profiles (very different in the case of
the herbaceous border); and this leads me on to the important subject
of vegetational succession which I shall discuss in the next chapter.

First, however, I must refer briefly to a group of organisms that I have
so far neglected and that I earlier described as 'lower' plants; the algae,

mosses, liverworts, ferns and horsetails, together with fungi and other micro-organisms. These too have an important role to play in a plant community and some have very special significance when that community is a garden. Although, collectively, this is a very diverse group, its members can conveniently be placed in the loose divisions shown in Table 6. None of them produces flowers or seeds: most produce much simpler bodies, known as spores, for reproduction and dissemination. The 'green groups' (algae, mosses, liverworts, ferns and horsetails) all perform photosynthesis in much the same way as more conventional plants, although structurally they are quite distinct, lacking true roots and, in the cases of algae, mosses and liverworts, lacking also any sophisticated conducting tissue. With the exception of horsetails and a few ferns, all are dependent on a very moist environment to survive, for they have no thick, protective 'skin' to help in water-retention. Indeed, most algae are aquatic (blanket weed in the garden pool and seaweed are familiar examples). Most take a relatively inconspicuous and insignificant role in most plant communities (although there are marked exceptions, such as the intertidal sea-shore zone where algae are actually the most important form of life). Algae, mosses, liverworts and ferns can be found in most of the layers or strata within terrestrial communities, and many are epiphytes, growing on trees or other perennials. Horsetails are an ancient group, and their ancestors adopted tree form in the coal swamps of the Carboniferous period. They impose their presence in gardens by virtue of being almost ineradicable weeds.

Fungi are a quite distinct, although diverse assemblage, all lacking chlorophyll and obtaining their nutrients from other organisms, plant or animal, living or dead. For most of them the basic structural units are hyphae, microscopic, tubular threads that aggregate to form a mycelium and, *en masse*, are responsible for the enormous range in form from the mould that grows on damp walls and the mildew on your roses, to the toadstools that either decorate or disfigure (depending on your point of view) the lawn. Fungi are ubiquitous members of every ecological community, and they play a vital role in the processes of decay that ensure the recycling of nutrients. While these so-called saprobic fungi remain largely unseen in the garden, the parasitic forms, which feed on living plant tissue and bring about disease, are more evident; it is often attack by native species of parasitic fungi on exotic garden plants that frustrates the gardener's desire to achieve ecological harmony.

Although presumably no-one gardens in the inter-tidal zone of the sea shore, it is an interesting habitat for any gardener wishing to obtain a complete picture of British vegetation types, for it is the only habitat dominated by large forms of algae. These seaweeds are related to blanket weed in the garden pool, and to the green growths on tree trunks.

Somewhat similar to fungi in their ecological role are bacteria, although they are structurally much simpler, wholly microscopic in size, and tend to be more important as saprophytes than parasites in a plant community. But as causes of diseases such as fire blight of rosaceous trees, and soft rotting of fleshy stems and tubers, their role in this regard is scarcely insignificant.

Viruses have no claim to be mentioned in an account of the classification of plants, for they occupy a quite unique place among living things: in many respects they are more like self-replicating chemicals than true organisms. They are all parasitic and depend for their continued survival on usurping some of the cellular functions of the organisms they infect. Their ecological role, therefore, is very much that of causing the demise of both plants and animals, and is most manifest in the occasional epidemics that bring a swift check to any species that might otherwise swamp a particular habitat.

Chapter Five

Natural vegetation and the garden

By definition, a garden is a collection of plants growing where they do because of man's influence. It is not a unique environment by any means; farms, sports grounds and public parks immediately spring to mind as other examples of sites where the vegetation is the product of human intervention. In fact, within the British Isles there are almost no areas of vegetation in the condition in which they would be if we or our ancestors had not, directly or indirectly, performed some particular action. Ask most people to name an unaltered British habitat and they might proffer a Scottish moorland, a Welsh bog or a Cornish cliff. Indeed, standing in the centre of a Highland moor with heaths and heather beneath your feet, it is hard to imagine the habitat ever having appeared different. Yet different it may well have been: the part-decomposed tree stumps that protrude from the peat are testimony to the former existence of the great Caledonian forest of oak, Scots pine, birch, alder and other trees, a forest where brown bears, wolves and wild boar roamed, where beavers dammed the forest streams and golden eagles wheeled above. The trees have gone, burned or felled by our ancestors, who introduced the sheep, goat and rabbit to graze away the forests' attempts at regeneration; and so it is over all of our islands.

Establish a garden in Britain today and you will begin to cultivate soil that has experienced generations of human influence. Every gardener inherits both ancient and recent history in his garden, and to understand the lessons of the past is to appreciate the present and plan for the future. Before describing the vegetational history of Britain, however, it is important to elaborate on a concept that I introduced in the last chapter: the sequential changes that occur over a period of time in any plant community, and the factors that influence them.

Vegetational succession

There is, close to my home, a stretch of limestone hillside that has been grazed for centuries by sheep and rabbits. It has the appearance of other grazed hillsides, in the predominance of rosette and creeping plants among the life forms, their basal or sunken meristems preserving them from the animals' teeth. A few years ago, the owner of the land fenced off a small area and the sheep and rabbits were excluded. That area is now unrecognisable as the old grazed hillside. Small hawthorn bushes (*Crataegus* spp.) are everywhere, interspersed with ragwort (*Senecio jacobaea*), docks (*Rumex* spp.) and other species with the bushy life form that I described on p. 94. Although seeds of hawthorn and the many

The Scots pine, Pinus sylvestris is one of three native British conifers, but even these fine Perthshire trees are unlikely to be wild stock and are but pale shadows of the original Scottish strains.

other species now so obvious had been dropped or blown into this area of hillside for countless years, none had previously realised its potential for growth because the livestock browsed away their tops. With the removal of the grazing pressure, that stretch of land has been colonised by a whole range of different species.

By a very similar token, if you remove the pressure that the hoe and the weedkiller impose upon your garden, it too will take on a completely different aspect. Thus you can see that changes do take place as a result of your actions; but how can we find out about the changes that preceded our acquaintance with the land and, indeed, how are we to judge what changes may take place in the future?

To discover facts about the history of your garden, the most obvious course of action is to talk to the previous owners of the land, to ascertain if old photographs of the site exist or if details of any sort are given in the deeds of the property. If you are lucky, this might give you information extending back for a hundred years or so. Unless you are very fortunate, you will then draw a blank; but you will only have used human, documentary sources in your search. To the enlightened interpretative skills of the palaeobotanist, however, every plant habitat, garden, farm or bog has its own, inbuilt documentation that provides clues to the history of the site, and which may enable him to draw up a picture extending back many thousands of years. Many scientific techniques are now employed in the study of vegetational history, and the radiocarbon dating method may allow a fairly precise age to be assigned to a fragment of plant tissue. Nonetheless, the simplest methods are still often the most rewarding and evidence for what were the predominant species (as opposed to the age of their remains) are those extremely durable plant structures, wood, seeds and pollen.

While it is fairly easy to identify the species of a woody plant from microscopic examination of a small fragment of wood unearthed from deep down in the ground, a sequence of annual rings is even more revealing. Even if the plant died many years ago, the relative width of the rings and the size of their cells will indicate how growing conditions and climate varied in the past. Seeds and pollen, with their hard outer coats, can survive for long after they have lost their viability, and thus indicate the types of plant that grew in an area in ages past. Under the microscope, the characteristic differences between the pollen and seeds of different species can be seen.

Take a very simple example: if I dig a vertical hole, or take a precise core of soil with an auger, from my garden and dissect successive short lengths of the core, I will find seeds and other plant remains. Logic suggests that the further down the core a particular type of seed or pollen grain appears, the longer ago will it have dropped onto the ground. If I find wheat seeds some way below the surface, therefore, I may conclude that, at some time past, wheat grew on the site where my garden now stands. Unfortunately, the situation is greatly complicated by the fact that I and my predecessors have dug the soil and mixed up the different layers. Moreover, below a certain depth, I will very probably find no recognisable seeds or pollen grains in the soil at all, because like most garden soils it is replete with bacteria and other microorganisms that degrade even durable seed and pollen coats after a few years. But if a comparable core is removed from a mud deposit, such as the silt of a lake bed, or the peat of a bog, where in the poorly oxygenated conditions decomposition is minimal, a vast fund of information can be obtained about the plants of the surrounding area. From such analysis, a picture can be drawn of the changes in plant life that have occurred over many years and it becomes apparent that quite predictable sequences or successions of species take place in particular environmental conditions. While it is impossible to be certain of the successions in *all* sites, a particularly well understood example will serve to incidate the type of pattern that can occur.

In Diag. p. 111 are shown the changes that, over a long period of time can convert a pool into a wood. During this progression, the habitat changes from open water through marsh, bog and wet woodland, to drier woodland and, finally, fairly dry pinewood. As this occurs, one community of plants is gradually replaced by another, each plant species effecting change on its surrounding environment very slightly, but sufficiently to enable other species to flourish or to cause existing species to decline. The newly flourishing species too, in turn, alter the soil or microclimate so as to allow yet another species to colonise. And, ultimately, for every natural environment, a more or less stable community arises within which individual plants grow and die, but where the overall structure, appearance and species composition remains much the same. This is known as the climax community and, in any area, tends to be the most complex and to include the largest species that local soil and climatic conditions allow.

In any geographical region, the bulk of the area will be occupied by a climax community whose nature is influenced primarily by climate. Such a widespread community is known as a plant formation and the plants that occur most numerously in the most important layer of the formation are said to be the dominant species. Over most of the British Isles, the climax community is the European summer forest formation, dominated by deciduous trees, but it should not be imagined that such a formation is a homogeneous entity. It has been estimated that only about 60% of the land surface of Britain was once forest, for within such great climatic formations are smaller formations, influenced by local soil factors (the waterlogged sites bordering rivers, and the cliffs forever washed by the sea, for example) or by animal or human activity. And thus, with human activity, we return to the plant community of the garden where we are forever fighting the natural tendency of the site to resume a progression towards some climax community of the natural deciduous forest. We cannot arrest this ecological evolution except by our constant attention. Today we have a lawn; tomorrow we may stop mowing it, and by next year we shall have a hay meadow; in a few more years there will be scrub; and ultimately we shall bequeath to our successors a wood.

I have referred already to the fact that many of our garden plant species are alien or exotic, in the sense that they originate in other parts of the world. Many, therefore, originate in different ecological formations, and sit uneasily in our deciduous summer forest environment. The map on p. 113 shows the major climax plant formations of the world, and serves as a reminder that gardening is a highly unnatural process: we are forcing plants from many different environments to compete with each other on unequal terms. In some instances, perhaps many instances, it works surprisingly well; Himalayan birches and rhododendrons appear to live extremely harmoniously with a British oak wood, despite the wide disparity in the rainfall, temperature ranges, light intensity, seasons and other wild-life with that in their natural homes. Some species clearly have wide tolerances of conditions outside their optima. Others clearly do not, and the Siberian larch (*Larix sibirica*), for example, is quite unable to cope with our 'fits and starts' seasons which may expose it to a few mild days in January, trick its physiological processes into believing that spring has arrived, and result in leaf buds bursting only for them to be laid low by a return to frosty weather. I shall discuss the problems

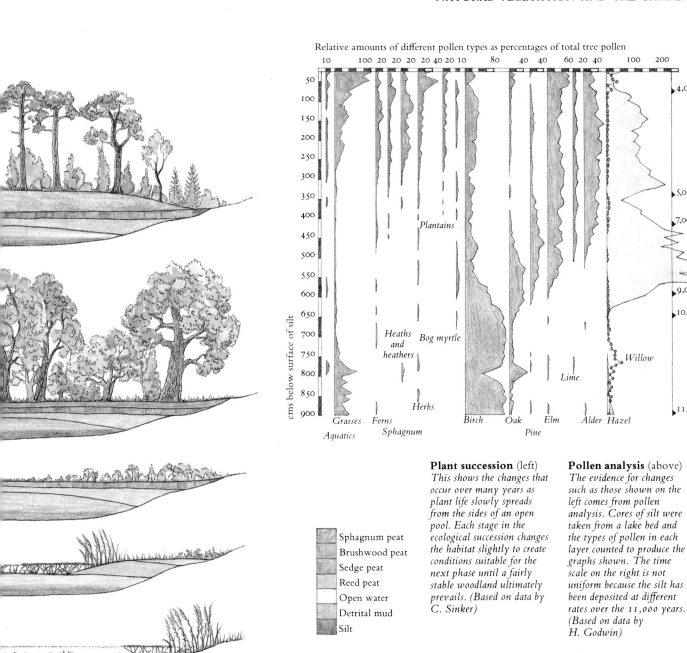

Relative amounts of different pollen types as percentages of total tree pollen

Grasses
Aquatics
Ferns
Sphagnum
Heaths and heathers
Bog myrtle
Herbs
Plantains
Birch
Pine
Oak
Elm
Lime
Alder
Hazel
Willow

cms below surface of silt

Number of years ago that the pollen was deposited

Plant succession (left)
This shows the changes that occur over many years as plant life slowly spreads from the sides of an open pool. Each stage in the ecological succession changes the habitat slightly to create conditions suitable for the next phase until a fairly stable woodland ultimately prevails. (Based on data by C. Sinker)

Pollen analysis (above)
The evidence for changes such as those shown on the left comes from pollen analysis. Cores of silt were taken from a lake bed and the types of pollen in each layer counted to produce the graphs shown. The time scale on the right is not uniform because the silt has been deposited at different rates over the 11,000 years. (Based on data by H. Godwin)

Sphagnum peat
Brushwood peat
Sedge peat
Reed peat
Open water
Detrital mud
Silt

III

associated with growing alien species in gardens later; but a little more should be said about the nature of competition between plants and of what makes a good competitor.

All plants are striving to enable their species to survive and, to achieve this, they must grow to maturity and reproduce. They can only do so if they have adequate light, air, moisture and nutrients for their own unique array of physiological processes to function efficiently. But all other plants in the surrounding environment are seeking the same ends and, clearly, there is a finite physical space within any plant community that satisfies the requirements of each species – often called its ecological niche.

If two species have the same or very similar requirements for environmental conditions, they will be in direct competition with each other, and any factor that can tip the balance away from one will be very important. This factor could be a pest or disease that suppresses its speed of growth, a slightly greater susceptibility to late frost, a higher seed production (which means that there will be numerically more individuals), or any of countless other small variables.

Within a single species, moreover, the natural variation generated by sexual reproduction as well as the chance production of mutants or 'sports' may result in individuals that differ in competitiveness from their parents. Most mutants are less successful – chlorophyll deficient variegated forms for instance tend to grow more slowly and are usually unable to compete satisfactorily. Some mutants appear to have equal advantages (the smooth skinned peach mutant called the nectarine, for instance) but positive evidence for natural or spontaneous mutations (as opposed to those induced artificially by plant breeders) having greater competitive advantages seems extremely scant.

It is because our native plants have, through thousands of years of competition, each 'found' their own ecological niches that they are so efficient at competing with the alien garden plants that we deposit in their environment. This is not to say, however, that all introduced or alien species are unsuccessful. Some of the most successful of present-day weeds were originally aliens; but many of them were introduced with crop species and thus found a ready-made artificial habitat, virtually identical to the one they came from. Many have also had a very long time to acclimatise; indeed, in the case of the field poppy (*Papaver rhoeas*), it is no longer possible to be certain whether it is native or not.

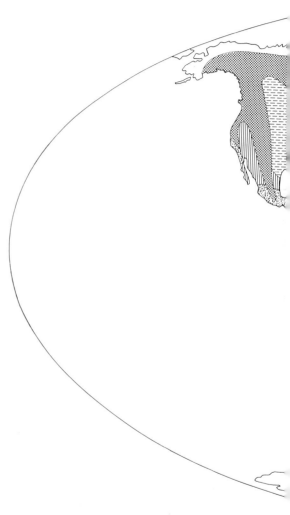

World vegetation
There are many ecological communities, but the world's plant life can be divided into broad vegetation zones. The British Isles are in the European Temperate Deciduous Forest with a small area of the Northern Coniferous Forest.

	Savannah		Tundra
	Coniferous forest		Desert and semi-desert
	Other forest and woodland		Grassland

Nor is there any telling how many species have attempted to colonise British habitats in the past, but have been unsuccessful.

Nonetheless, it is seldom that an exotic species, newly planted in any particular ecological niche, finds itself more suited to that niche than the 'local residents'. Thus it is that the weeds usually triumph in a garden if we don't intervene to suppress them. Only very occasionally does an artificially introduced plant, by chance, immediately prove more successful than the native species and, when this does happen, it is likely to oust the present incumbents. The very rapid spread of the Himalayan balsam (*Impatiens glandulifera*) along many English river banks is a well-known instance of this; and within gardens there are a few species that can so overwhelm the community as to seem to be turning on their owners: some of the creeping campanulas and mind-your-own-business (*Helxine soleirolii*) are among the best known.

The history of British vegetation

I have described the natural climax vegetation of much of Britain as typical of the European deciduous summer forest; in the North of Scotland it gives way to the north European coniferous forest. And yet, there is very little of either to be seen in Britain today. Where has it gone? – and what do the subsequent events mean for gardeners?

In fact the forests, over much of Britain, have gone not once but several times, the climax formations having been swept away in the Ice Ages and replaced temporarily by communities typical of the present day northern tundra. Consequently, at different times, many plant species have flourished in Britain that no longer exist here. More significantly, we have lost several major species that might be expected in our present-day native forest floras, but never recolonised after the ice finally retreated: Norway spruce (*Picea abies*), hemlock (*Tsuga* spp.) and silver fir (*Abies alba*) most notably, although it will interest gardeners to know that other species that once grew wild here include *Rhododendron ponticum*, Norway maple (*Acer platanoides*) and the grape-vine.

The last Ice Age ended between 10,000 and 15,000 years ago and the forests subsequently returned, only to disappear again at the hands of our ancestors. The key to this latest demise lies first in the nature of the dominant species, principally oak (*Quercus petraea* and *Q. robur*), beech (*Fagus sylvatica*), ash (*Fraxinus excelsior*), birch (*Betula* spp.), elm (*Ulmus* spp.), lime (*Tilia* spp.), alder (*Alnus glutinosa*) and hornbeam (*Carpinus betulus*) in the deciduous, and Scots pine (*Pinus sylvestris*) in the coniferous

Changes over the past 14,000 years in the area
of the British Isles covered by forest

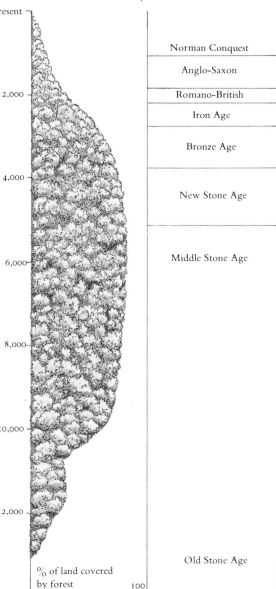

Present

2,000

4,000

6,000

8,000

10,000

12,000

14,000

% of land covered
by forest   100

Number of years ago

Norman Conquest

Anglo-Saxon

Romano-British

Iron Age

Bronze Age

New Stone Age

Middle Stone Age

Old Stone Age

forest. Such trees yield timber and for the past 5,000 years, men have felled them to build their homes, fortresses and ships and to fuel their fires. Second, wherever trees grow and whatever their species, their size prevents man from indulging in another of his characteristic activities, agriculture: the growing of crops and the rearing of animals. So forests were cleared to make land available for farms and, for the first time, alien species were deliberately imposed on British ecology. Analyses of pollen and seed in ancient peat and mud deposits show graphically the sudden occurrence of these aliens, the decrease of the trees, and, as an extremely significant consequence, the increasing presence of the many species that we now recognise as typical of disturbed ground (Diag. p. 111). These are the weeds of cultivation, many of them fast-growing and maturing annuals that compete very effectively for light, water and space with the less well-adapted crop species from alien ecological communities. A graphic example of the competitiveness of 'weeds' is that of the cereal rye, which seems to have originated in central Asia and may have occurred in Britain as a weed among crops of wheat, barley and oats. Realising that they were fighting a losing battle on poor soils and in cold climates, the northern farmers turned to the weed instead of the crop and began cultivating rye to the extent that by mediaeval times it assumed a predominant place in British agriculture. Only by dint of modern methods of crop husbandry and improved varieties has wheat ousted it.

The final blow to our natural vegetation was dealt by the gradual enclosure of the land that culminated at the end of the eighteenth century in the Enclosure Acts, which very demonstrably rendered unto the farmer that which was the farmer's, and enabled individuals to make their mark on the landscape in a particularly enduring way. As in all manners of human interference with nature, however, there was even here some benefit: it was derived from the creation of a novel habitat in the shape of the hedges and, to a lesser extent, the walls that provided the field boundaries. Indeed, while the present appearance of most of the British landscape is the result of the replacement of the natural climax plant formations with communities shaped by human influence, these do allow a gardener to see in miniature a range of habitats and different sub-climax communities.

Present-day British vegetation

Arable land, market gardens, orchard and grassland

One of the best ways to appreciate the range of plant communities in modern Britain is by reference to a land-use map (Map p. 119). Although the farm cropping data will obviously change from year to year, the major categories of arable land, market garden, orchard, grassland, woodland, heath and moorland, water, marshland and coastal habitats (in addition, of course, to industrial and other built-over areas) will be fairly constant. What cannot be shown on a small-scale land-use map, however, are private gardens which individually are too small, but which in total cover over 3,000 square kilometres.

For these first four of the land-use categories 'existing vegetation' is a more appropriate term than 'natural vegetation'. Collectively, they now occupy almost 190,000 square kilometres, an area slowly being depleted as housing and roads make increasing demands on the available space. Thus, private gardens are continually being created on land with a long history of growing crops; and even more gardens are surrounded by such land with its attendant pollen, seeds and weeds. Interestingly, very few of the crops themselves have become so well adapted to our environment as to be listed officially as introduced components of the British Flora and although I shall be considering the origin and ecology of some of them in the next chapter, there are certain features of such previously cropped land that are best mentioned here.

The largest component crops of British agriculture are members of the grass family: cereals, and grass for grazing, hay and silage. Next in importance are potatoes, sugar beet, fodder crops such as swedes, turnips and kales, and the most recent introduction, oil-seed rape. The remaining 2,500 square kilometres are occupied by horticultural crops and, although this means principally vegetables and fruit, it is not synonymous with the old categories of market-garden and orchard, for three quarters of our national vegetable production is now based on arable farms rather than small-holdings. (The vegetable crop grown most extensively, by a long way, is peas, although their production is concentrated in a restricted area of eastern England.) Your garden, if derived from old vegetable growing land, is almost certain once to have borne brassicas, collectively the most important of our vegetables, and probably beans, bulb onions and lettuces too. Extensive areas of orchard occur principally in Kent, the Vale of Evesham and Herefordshire, with apples as easily the largest crop. And one of the most conspicuous declines among

Table 8 Efficiency of some common annual weeds in respect of seed production and survival

Species	Average number of seeds produced per plant	Approximate longevity of seeds in soil [years]
Aethusa cynapium (fool's parsley)	6,000	10
Capsella bursa-pastoris (shepherd's purse)	4,000	30
Cardamine hirsuta (hairy bitter-cress)	600	?
Chenopodium album (fat hen)	3,000	30
Euphorbia peplus (petty spurge)	250	?
Fumaria officinalis (fumitory)	few	30
Galium aparine (cleavers)	few	?
Lapsana communis (nipplewort)	1,000	?
Matricaria matricarioides (rayless mayweed)	7,000	?
Papaver rhoeas (field poppy)	17,000	80
Senecio vulgaris (groundsel)	1,000	?
Solanum nigrum (black nightshade)	10,000	40
Sonchus spp. (sow thistles)	18,000	?
Stellaria media (chickweed)	2,500	?
Urtica urens (annual nettle)	1,000	?
Veronica spp. (annual speedwells)	2,000	10

Table 9 Some evidence for 'one year's seeding' resulting in 'seven years' weeding'

(A stock of weed seeds was artificially mixed with a quantity of garden soil. All seedlings that emerged were removed before they flowered so no additional seeds could be added. At yearly intervals, the soil was analysed and the number of viable seeds recorded as a percentage of the original total. It was found that, after seven years, there were less than 1% of the original total remaining.)

Number of years after the stock of weed seeds was placed experimentally in garden soil	Percentage of viable weed seeds recovered each year
1	47
2	27
3	12
4	5
5	4
6	3
7	1.5
8	0.5
9	0.5
10	0.5

Changes in the field boundaries of a Herefordshire parish over 200 years

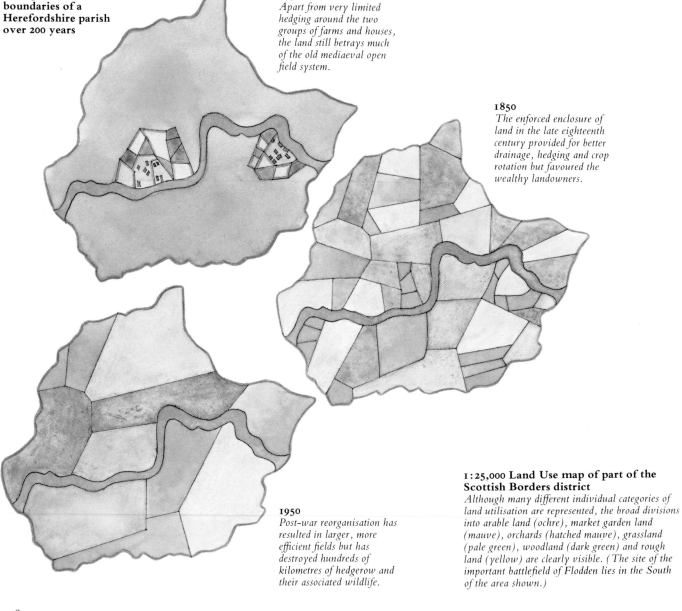

1750
Apart from very limited hedging around the two groups of farms and houses, the land still betrays much of the old mediaeval open field system.

1850
The enforced enclosure of land in the late eighteenth century provided for better drainage, hedging and crop rotation but favoured the wealthy landowners.

1950
Post-war reorganisation has resulted in larger, more efficient fields but has destroyed hundreds of kilometres of hedgerow and their associated wildlife.

1:25,000 Land Use map of part of the Scottish Borders district
Although many different individual categories of land utilisation are represented, the broad divisions into arable land (ochre), market garden land (mauve), orchards (hatched mauve), grassland (pale green), woodland (dark green) and rough land (yellow) are clearly visible. (The site of the important battlefield of Flodden lies in the South of the area shown.)

our crops in recent times has been that of cherries, and there is a good chance that a new garden in Kent lies on the site of a former cherry orchard.

Almost all farm and horticultural crops are grown as monocultures and the community of a field of oil-seed rape, wheat or Brussels sprouts is probably as far removed from any natural community as is possible. Even so, such monocultures can teach gardeners certain ecological principles. The impact of pests and diseases will be very evident in almost any field where the farmer has neglected to take control measures and the rapidity with which such problems can spread when no natural barriers are present is often self-evident. It is easy too to see the effect that competition with weeds can have when the weedkillers on which modern farming is so dependent have been applied too late, incorrectly or not at all. Oil-seed rape, the crop that has had an almost explosive impact on British farming in the past fifteen years, displays the ability to invade the existing environment. Roadside verges and hedgerows in many parts of the country are now coloured almost as yellow as the fields at flowering time. Notably, this is a crop not altogether alien, for its close relatives and some of its ancestors are either native or very well-established members of our flora; I shall consider the possible significance of this in the next chapter. Oil-seed rape makes an interesting contrast, however, with most cereals, which rarely spread far beyond the immediate field boundaries and appear much less competitive. Very few crop plants indeed have become so well established in Britain as to occur as integrated members of the flora. Partly this may be, as with apples and other fruit trees, because they are hybrids and don't 'come true' from seed; but generally poor competitiveness with native vegetation is clearly a major factor. For the gardener, this means that any crop plants that occur inexplicably in his garden are most unlikely to be relicts from very old cultivation; even fruit trees will be the old, planted individuals themselves rather than their progeny.

Woodland

Do not be misled by the extensive tracts of woodland that you will find in some parts of the country. Although there are something over 2 million hectares of trees in Britain, most of them represent not the surviving parts of the ancient climax formation, but forest and wood artificially planted either for timber or game cover. To see woodland that is believed to be truly relict, you will need to visit such remote areas

The scenery of much of lowland Britain, in its fairly small fields, hedges and woods betrays many centuries of enclosure and land use.

as the Black Wood of Rannoch, as representative of the northern forest, or Wistman's Wood on Dartmoor which is a depauperate form of old oak forest. The extensive areas of coniferous forest in Britain today are very largely the results of planting by the Forestry Commission and by private landowners in this century. The species they contain are alien, most notably Norway and Sitka spruces (*Picea abies* and *P. sitchensis*), Western red cedar (*Thuja plicata*), larches (*Larix* spp.), and various pines (*Pinus* spp.) – and even the planted Scots pine bears little resemblance to the majestic native tree. These dark, often forbidding forests have been called ecologically sterile. This is a contradiction in terms. In fact, the bird life of some such forests is surprisingly rich; but, certainly, a monoculture of Sitka spruce makes a formidable competitor for other plants, and casts a year-round shade so dense that almost nothing else will grow beneath it. Only with the regular felling for timber, every thirty years or so, does the land burst forth again with lower growing plant life-forms, as long-dormant seeds germinate and others are blown in from neighbouring communities. Even a tree as potentially massive as a Sitka spruce cannot compete on equal terms with bracken, native grasses and other herbaceous plants in its early growth; and it is the mark of the good forester that he pays assiduous attention to weed control for the first few years after replanting a cleared site. It is often forgotten by gardeners that a young tree is no larger than many herbaceous species and, when it is an alien to boot, will have to fight for its life.

There has been much less planting of deciduous than of coniferous forest, for deciduous trees usually grow much more slowly and yield useful timber only after many more years – timber which is uneconomic for the paper pulp market that now devours so many of our conifers. In the past, extensive deciduous plantings were made on private estates and royal parks to provide house- and ship-building timber and firewood, and to provide game cover. The two most notable survivors of the old plantings (although now greatly altered and containing many conifers too) are the New Forest in Hampshire and the Forest of Dean in Gloucestershire. All these forests were artificially managed, moreover, as some trees were removed, others planted, the shrub and field layers of vegetation suppressed by cutting and introduced deer encouraged to browse. Even in their early years, therefore, they were not dependent on the natural death and regeneration that is the hallmark of a natural climax community.

This Warwickshire field reveals much of modern British farming in its limited cereal monoculture and retention of some hedgerows and trees (although the latter greatly diminished by Dutch elm disease over much of the country).

More recently, there has been apparent a tendency towards more planting of mixed deciduous and coniferous forest and woodland, largely for reasons of aesthetics and amenity. This has been generally welcomed by ecologists as at least giving an opportunity for a rather greater proportion of native species to survive than is the case in the purely coniferous stand, while providing a better financial return to the owner than the purely deciduous planting.

Several rather specially managed types of woodland must be mentioned as creating unusually interesting ecological communities. In the southern half of England, most notably, will be found deciduous woods of oak, hazel (*Corylus avellana*), sweet chestnut (*Castanea sativa*), hornbeam, birch, alder and ash in particular, in which the trunks arise in massed groups from a single base. These are the outgrown results of coppicing, a once widespread practice that was an attempt to obtain hardwood timber rather more quickly than waiting for trees to mature. A tree was felled and the cluster of small shoots that sprung up from the stump allowed to develop. Every ten to twenty years, these were cut to provide poles for fencing or other uses. Sometimes, in the coppice-with-standards system, some of the trees were allowed to grow to full height to provide larger timber. Pollarding, the regular cutting back of the crowns of individual trees, is also based on the ability of some species to regenerate shoots. It doesn't work with conifers, which lack this ability.

Both coppicing and pollarding are among the most functional yet degrading uses to which a tree can be put, and the results are unsightly; witness the pollarded limes of many of our urban streets which should teach the gardener a lesson in species selection. If you cannot accommodate a full-grown tree of any species in your available space, then choose a smaller species rather than hideously abbreviate a large one. Nonetheless, a functioning coppice (and a few are still managed as such) does permit the observation, over a fairly short time-scale, of the process of colonization of a former woodland floor by herbaceous plants, but with the significant difference that the tree growth originates not with seedlings but with a mature stump and root system, and can always therefore compete very effectively.

Finally to perhaps the most interesting, if most obviously artificial, sub-climax woodland community, the hedgerow. A hedge, in very simple terms, can be thought of as a long, narrow wood subject to fairly rigorous management. The plant community comprising the hedgerow is normally dominated by the species planted to form the basis of the boundary. Hedges are usually planted with only one species of tree or shrub, but others will colonise naturally as seeds are blown in or carried by birds or animals. The hedgerow equivalents of the field and ground layers are always the result of natural colonisation, and it is the field layer that is subject to periodic clearance in an intensively managed hedge. The total number of plant species recorded in British hedgerows is very large, and as many as three hundred are quite common in this country, to the extent that the English names of some plants have hedge as a prefix; hedge garlic (*Alliaria petiolata*) and hedge mustard (*Sisymbrium officinale*), for instance. While there are some very marked regional variations in the species composition of hedgerows, reflecting climatic or historical factors (fuchsia in the south-west of Britain, or beech around Exmoor, for instance), there are also clear national trends. The commonest tree species are oak, ash, beech and sycamore (*Acer pseudoplatanus*). (Elm, formerly of major importance, has declined dramatically as a result of Dutch elm disease.) Sometimes the trees of hedgerows are the result of individual specimens of the planted hedgerow species having been allowed to grow to maturity; sometimes they have been planted in addition to the main hedgerow species; frequently they have colonised naturally and been allowed to grow to maturity. The plant used most commonly for farm hedgerow planting is hawthorn, although in gardens, the Japanese privet (*Ligustrum ovalifolium*) is easily the most widespread. Most other shrubs that now occur so commonly in hedgerows have colonised naturally; such plants as wild privet (*Ligustrum vulgare*), dog rose (*Rosa canina*) or elder (*Sambucus nigra*) are most unlikely to have been planted as farm boundaries as, on their own, they are not robust enough to contain livestock.

From extensive surveys of the species composition of hedgerows and careful study of old maps and recorded planting dates, it has been possible to devise a system for estimating the age of hedges. Given a number of qualifications, such as the need for several samples, the need to avoid lengths of hedge that adjoin woods or gardens (where, clearly, deliberate planting is a major influence), a rough rule of thumb is that, within a 30 metre run of hedgerow, the number of species of tree or shrub, counted from one side, indicates the age of the hedge in hundreds of years. I have a length of garden boundary hedge containing hawthorn, yew, ash, holly (*Ilex aquifolium*), wild gooseberry (*Ribes uva-crispa*), bramble

Young Scots pine planted among the heathers beneath birch on a New Forest heath.

A beech and sweet chestnut wood in July, its dense shade suppressing the ground flora.

126

Wet birch woodland, with a sparse field layer and with a ground flora dominated by Sphagnum.

(*Rubus* spp.) and blackthorn (*Prunus spinosa*). While this cannot give me a precise date for my hedgerow, I would be prepared to bet that the large number of wild species means that it is probably the original boundary of the property (the house is about 350 years old) or even an earlier field hedge incorporated into the boundary.

I shall be discussing the special significance of hedgerows in garden ecology later, but all hedges provide a 'pocket' example of a layered community, with dominant species and two or more under-storeys, and display the effects on plants of shade, severely restricted water penetration and nutrient depletion.

Heath and Moorland

A heath or moorland is a plant community largely shaped by the acidity of the soil, with extensive growth of very small shrubs and bushy plants, particularly members of the family Ericaceae – heather (*Calluna vulgaris*) and heaths (*Erica* spp.). It may be the climax community of high, exposed areas such as some of the northern mountains, or a sub-climax community, as in the south of England where, with the removal of such pressures as grazing by rabbits, deliberate cutting or fire, a mixed woodland of birch and pine would probably establish itself.

There are lessons here for the gardener in his choice of ground cover species, for tree seedlings quite commonly fail to survive even on undisturbed heath, because of the effectiveness of the ericaceous plants in limiting the penetration of light and water, and possibly because the mycorrhizal fungi needed for the establishment of some tree seedlings are inhibited by substances present in heather peat.

Heaths and heathers are not, however, the only plants of the heath and moorland communities. Some sedges, especially the misleadingly named deer-grass (*Scirpus caespitosus*) and cotton-grasses (*Eriophorum* spp.), are common and may be locally dominant. The bilberry (*Vaccinium myrtillus*) can dominate extensive areas of moor, usually at higher altitude than heather as it is more tolerant of exposure (although it is, interestingly, more shade-tolerant too; so it also occurs at lower levels in the field layer of birch and other woods on acid soils). In more than almost any other British plant habitat, the lower plants, especially mosses and lichens, are of major importance in the ground layer of moorland.

In the less well-drained areas of moorland, bogs are common, with their own characteristic vegetation, which I shall discuss in the following section with other wetland habitats.

For convenience, I shall stretch this account of moorlands to include mountain-top or 'Arctic–Alpine' habitats – not because they are especially similar, but because the former often grades into the latter at altitude. I know of no garden actually situated on an exposed mountain top, but the plants of this environment (or at least of comparable, although even higher, environments in other parts of the world) make up much of the flora of the alpine garden. That the growing of many alpine species is so often beset with problems is because of the great difficulty in emulating the conditions of the natural mountain top. In saying this, I appreciate that the mountain top itself is far from a uniform community, but it is one always shaped by extreme paucity of soil and by extreme climatic conditions. Rainfall is high, but, because of the nature of the underlying soil and rock, drains away very rapidly. Temperatures fluctuate wildly, from long periods below freezing, to the almost unbearably hot; try placing your hand on the surface of an exposed mountain-top rock at noon on a clear summer day to appreciate what low-growing alpine plants must tolerate. One of the most unusual environments within the mountain top community, and one that is especially satisfying to recreate in a garden, is a scree, a collection of loose stones and boulders in a more or less unstable condition. Such a habitat has a minimal soil content, and very free drainage. Because of the drying effect and sheer physical strength of the wind, the near-absence of soil, and the difficulty for pollinating insects of flying more than a few centimetres above ground level, the climax community comprises the life-forms of cushion, rosette and creeping plants or very small shrubs – all with their survival buds well protected.

Wetlands   A wetland is any habitat where impeded drainage and/or a constantly replenished supply of water (either directly, as high rainfall, or indirectly, because of the proximity of a watercourse) ensures that the land hardly ever dries out. The most important of these communities in Britain are the various types of bog on very acid peat soils, fenland on less acid peats, and marshland on silty soils, together with the more local versions of them that occur on river and lake-sides.

The community called blanket bog occurs in place of moorland where rainfall is very high and drainage impeded. Some plants are common to both types of community, but heather and the heaths cannot tolerate the saturated and particularly nutrient-deficient soil. Indeed, it is in these impoverished conditions that one group of plants, the insecti-vorous species, have evolved the habit of being able to digest insects that are trapped on their leaves, in order to obtain additional nutrition. Sundews (*Drosera* spp.) are the best known British examples. Mosses, especially species of *Sphagnum*, are often the dominant plants in extensive areas of blanket bog. Where *Sphagnum* colonises the remains of fen vegetation, a bog can form over more alkaline soils and is then called a raised bog. On p. 109, I described the process of raised bog formation as a part of the successional colonisation of open water that may lead, eventually, to a dry woodland.

Fenland is best seen in a few surviving areas in East Anglia, where it has been formed on the upper parts of old river estuaries. The dominant plants, and those whose remains contribute to fen peat (which is much less acidic than the *Sphagnum* or moss peats), are all of the life-form called mud plants; the great fen sedge (*Cladium mariscus*), the common reed (*Phragmites australis*), the blunt-flowered rush (*Juncus subnodulosus*) and reed canary-grass (*Phalaris arundinacea*).

The vegetation of marshes is essentially similar to that either of bogs or fens, depending on the relative acidity of the underlying silt.

Within the open water, in rivers and lakes, two main types of plant may be found. Floating plants, such as duckweed (*Lemna* spp.), have no anchoring roots and obtain all their mineral nutrients from the water. The true water plants, like water lilies (*Nymphaea* spp.) and milfoils (*Myriophyllum* spp.) are rooted in mud and have no aerial shoots, al-though the depth of water they can grow in varies greatly from species to species; a fact to be borne in mind in the planting and management of pools in gardens. A few plants of this type, such as water crowfoot (*Ranunculus aquatilis*) actually thrive in fast flowing water.

I shall mention the sea coast only in passing, for although many people do garden virtually on the shore-line, the communities are essentially the shrub, grassland or other types described already, with the difference that the component species are those able to tolerate exposure not only to wind, but more importantly to salt-laden air. I know of no-one, however, who gardens in the intertidal zone that represents one of the very few British plant communities dominated by algae!

The vegetation at a lakeside can give inspiration to a gardener through its richness in both species and life forms. The bog garden can mimic this on a smaller scale and incorporate choice exotic species too.

Managing the natural vegetation

Your garden may once have been almost any of the aforementioned communities; indeed, at least part of that community may remain in or may surround your property. I shall consider, next, therefore, how the previous history of the site might affect your management of the garden and the implications that it may have on your choice of species.

If I succeed in promoting only one message in this book, I hope it will be the notion that gardening need not, should not, be simply a matter of eradicating all that seeks to live naturally on a site in favour of that which you wish to impose in its stead. There are three aspects of natural vegetation which must be considered in relation to gardening: first, the extent to which plants already existing on a site can be incorporated into a garden; second, the extent to which wild plants can be used in the garden, even if they don't happen to be growing there naturally; and, third, the way in which to manage those wild plants that you certainly don't want, the weeds, but which will be in competition with the plants you *do* want. In this account, I mean by 'natural' the species of native British plants already existing on the site of a garden, even though, as with many of the species of hedgerows, some of them may have been planted deliberately at some time past.

The use of existing wild plants in a garden

By and large, for reasons that I shall outline in the next section, I don't believe that wild herbaceous plants can be very effectively incorporated in a well managed garden if they are simply left to their own devices where they have grown naturally. The one exception, possibly, is grass. To many gardeners grass is grass, and it doesn't occur to them that there are different grass species, just as there are of any other flowering plants. Indeed, the grass family is one of the largest of all: even in Britain there are well over one hundred different species, and many non-native species and hybrids are also used for lawns and other cultivated purposes. All grasses share the growth habit centred on the basal meristem that I described on p. 81, and they are therefore all theoretically amenable to being mowed. Can they all, therefore, be used to create a lawn without the need to resort to sowing or turfing with a selected blend of deliberately chosen varieties? In theory, yes, they can, largely because any community of grass that is likely to be the precursor of your garden will almost certainly have been sown by someone, probably a farmer. Nonetheless, grass sown for feeding cattle doesn't make good lawn turf; many of the grasses of which it is composed are coarse leaved, with a sprawling rather than neat, compact habit, tall growing, prone to flower and, almost certainly, occur in association with a considerable population of weed species. Because there will be a mixture of grasses of varying habit, moreover, any lawn created from them will be very patchy in appearance. It is much, much better, to start from scratch with turf or seed, and incorporate any existing rough grass, if you can, into orchards or other less refined areas of the garden. But a firm word of caution must be given about buying turf.

There are three main types of turf and you will see these designations used in advertisements. Sea-washed turf or Cumberland sea-washed turf is the finest of all turf. Originally, it was all dug from the Cumberland coast but the term sea-washed now applies to any turf from coastal saltmarshes. It contains a quite unique blend of wild grasses, mainly varieties of creeping fescue (*Festuca rubra*) and fiorin (*Agrostis stolonifera*) and is a very difficult turf to manage correctly, for its natural soil is silt rich in sodium. It is the turf of bowling greens; but it is not a suitable choice for gardens.

Downland turf (or, at least, much of the best downland turf), originates on the South Downs where a thin soil overlies chalk. The principal grasses that it contains are sheep's fescue (*Festuca ovina*), bent grasses (*Agrostis* spp.) and crested dog's tail (*Cynosurus cristatus*), and there may be greater or lesser amounts of clovers and trefoils. Downland turf is the turf of cricket squares and high quality lawns, and really should be chosen for a respectable lawn.

The final category is the one that you will see most commonly, meadow turf. People's opinions of what constitutes a meadow vary considerably and, in all honesty, meadow turf is usually the cattle grazing mixture that I described above. It should never be used for a fine lawn. Even the designation 'treated with weedkiller' that is often applied to it should be viewed with circumspection; it is certainly no guarantee that any of the weeds has actually been killed. I shall say more in the next chapter about the selection of the ideal cultivated grass species to use for sowing a lawn.

Assuming that you decide, for whatever reason, that the old grass must be removed (or, to be more precise, turned under *in situ*), you will be faced with two main problems. The first is a residue of soil-inhabiting animal life that had previously been very contentedly feeding on grass roots, and which will find your newly planted vegetables and flowers

An old Teesdale meadow in the height of summer; the richness of its species is astonishing to those more used to seeing grassland that is the result of intensive herbicide usage and that comprises a few, hybrid and alien grasses.

at least as palatable. Organisms that fall into this category are leather-jackets, the larvae of craneflies, and wireworms, the larvae of click beetles. I shall be discussing their role in natural plant communities and gardens later. The second major problem will be that of nutrient depletion and although different in nature and extent from the depletion that follows the removal of woodland, the reasons for it are essentially the same (see below).

Trees, shrubs and hedgerows, simply by virtue of their size, will already be playing a relatively dominant role in the existing plant community; if removed, they leave an appreciable gap. Some species of woody plant make much greater demands on their environment than others; generally speaking, those with a relatively shallow spreading root system will have the greatest impact below ground, for they will tap the nutrient and water resources of the soil over a wide area. If this is combined with a large and spreading crown that casts a dense shade (especially if this is evergreen), the impact will be greater still. Thus, among large trees, yew and beech, and among smaller trees, privet, create the largest area over which it is difficult to grow other plants. Indeed, it is sometimes thought that such plants actually poison the soil, so effective is their suppression of other vegetation. But there are few species for which the production of toxins of this type (a phenomenon known as allelopathy) has actually been proved; among trees, walnut (*Juglans regia*) is the best known. It produces a chemical called juglone in its leaves and, when this is washed into the soil, it can suppress the growth of some other species.

Thus, removing a beech, yew, privet or other large tree or shrub ought to give you the greatest benefit in respect of additional growing space. But remember that the removal of a large tree will also remove a great deal of wind protection and can totally change the ecological character of a site. It may also, incidentally, remove much of your privacy. There is the additional consideration, applicable to any tree, that it may enjoy legal protection*. It must be said also that the removal of large trees is not a task to be undertaken lightly, and really is a job for the expert. If you plan to use the space occupied by the tree for vegetables or fruit, the stump must be removed in its entirety – also a task for the expert, and much easier when a good length of trunk can be left to provide leverage. Sometimes stump removal may prove impossible and, very commonly, a gardener inherits stumps as a legacy of the clearance of woodland before his house was built. The stump itself can be made into a habitat for climbing and scrambling plants (see the next chapter); but this highlights a major factor that must be considered when a garden is created from old woodland, for the cut surfaces of hardwood tree stumps provide the means for the entry into the soil of a fungus called *Armillaria mellea*, known popularly as honey fungus. It must be admitted that the soil of any old deciduous woodland stands a good chance of containing the mycelium of this fungus anyway, and unfortunately it is one of the most efficient killers of plants, especially woody plants. On p. 139, I make some general comments about the control of fungus diseases, but honey fungus really is a case apart, for there is no effective chemical protection from it for garden plants. The fungus spreads through the soil by aggregated strands of mycelium called rhizomorphs, and it is sometimes possible to arrest their spread by sinking a physical barrier such as plastic sheet, or by drenching the soil with certain proprietary soil disinfectants.

Individual trees may be subject to a Tree Preservation Order, taken out by a local authority whose permission will be needed before it can be felled, lopped or otherwise interfered with. If your garden and the tree lie in a designated Conservation Area, moreover, it may well have legal protection automatically if it is over a certain size. In any case of doubt, you are strongly advised to check with your local authority before taking any action.

Another effect that an old woodland site bestows on a garden is what is usually described in gardening books as 'a highly impoverished soil'. This statement must be placed in its ecological context, for if a soil is truly impoverished, how has a flourishing woodland survived on it for so long? Behind the answer lies the whole intriguing nature of the life and death of plant communities and the meaning of the expression 'ecological niche'. The woodland has survived because the trees' roots have continually been tapping new, deeper and more distant regions of the soil, but have, in the process, certainly depleted other parts of some nutrients. Within this nutrient-depleted, shady environment have become established the species of the shrub, field and ground layers that not only tolerate shade, but have quite different nutritional demands from the trees. This is not, however, the ecological niche of vegetables, fruit or most ornamental garden plants, and the soil will, most assuredly, be deficient in the nutrients they require. A very considerable input of artificial fertiliser will be needed, therefore, to redress the inroads that possibly many hundreds of years of woodland growth have made.

Let us suppose, however, that your inherited trees cannot be removed, or that you simply don't feel inclined to remove them, and/or that growing fruit and vegetables doesn't interest you. Here is an opportunity to graft your gardening way into an established community rather than try to replace it with another. By selecting plant species from other, similar but exotic woodland environments, you can develop a woodland or shade garden with all the semblance of it being a natural creation, were it not for the fact that the component plants come from all corners of the globe.

Much of what I have said about existing woodland applies equally to orchards, although here there are additional factors to consider with regard to the retention or removal of old fruit trees. Most of the features of old market gardens and arable land are peculiar to cultivated vegetation, and they will be discussed in the next chapter.

If you have areas of natural heathland or wetland in your garden, you may consider yourself either blessed or cursed, depending on your gardening aspirations. Heathland means, above all, an acidic soil with a low nutritional status that renders it unsuitable for vegetable and fruit gardening. It is, for me, one of the classic sites where you should follow the old adage and 'go with your soil', for it is difficult if not impossible to fight this particular ecological community; to attempt to change it significantly is futile. Given such a site, I would examine the floras of the heathlands and acidic soils of the world, especially those of the Mediterranean region, the foothills of the Himalayas, and South Africa. These would yield many glorious species of *Calluna, Erica* and *Rhododendron* that, given a selection for acceptable hardiness, would blend happily with much of the existing native vegetation.

If your wetland is a bog or marsh formed as a result of some locally impeded drainage, it may be possible to instal an artificial drainage system. The resulting soil should be amenable, given intensive management and abundant fertiliser and lime, to being converted to fruit, vegetable and ornamental plant growing. Nonetheless, it must be said that a natural bog garden does afford splendid opportunities for growing a range of plants that otherwise are very difficult to cultivate; a collection of bog primulas, for instance, will surely reveal to its owner that a naturally boggy site, an ecological habitat so difficult to create artificially, can be a heaven-sent asset. River and stream courses are rather different matters, for you will be legally restricted in the extent to which you can interfere with the natural flow of water.

Managing the natural vegetation – less straightforward considerations

So far, I have discussed the existing natural or semi-natural vegetation of a site solely in respect of incorporating it into the garden for what it is. There is another important feature to consider, however, and this is one that I hinted at when I referred to the soil-inhabiting insect larvae beneath old grassland. The existing vegetation has its own associated plant and animal life and may well merit conserving for their sakes or, conversely, represents a refuge for organisms that could threaten garden plants. Oaks are the classic examples of plants that serve as a haven for wildlife; several thousand species of insect alone are associated with or dependent on oak trees. Increased interest in wild plants for gardens, and in wildlife conservation in general, has brought the importance of many more species of plant to the public's attention. Table 10 lists some of the more important wild plants that might be considered as garden species from this point of view. There are some interesting contrasts to these plants, however, in species such as sycamore which have a particularly depauperate associated flora and fauna. It is often suggested that the number of dependent species reflects the length of time that the 'host' plant has been present in Britain; sycamore, for instance is not native and is reckoned to have been introduced about 400 years ago. The

Table 10

Some wild plant species of special value in gardens as refuges for wild life*

Species	Special value
Beech (*Fagus sylvatica*)	Harbours many species of insect; fruits provide food for many birds and small mammals
Birches (*Betula* spp.)	Second to oak in importance as habitat for insects
Dead-nettles (*Lamium* spp.)	Highly attractive to bees
Elderberry (*Sambucus nigra*)	Fruits are valuable bird food
Foxglove (*Digitalis purpurea*)	Highly attractive to bees
Hawthorn (*Crataegus* spp.)	Fruits provide valuable bird bood; a habitat for many species of insect
Hemp agrimony (*Eupatorium cannabinum*)	Highly attractive to many butterflies
Honeysuckle (*Lonicera periclymenum*)	Highly attractive to many moths
Ivy (*Hedera helix*)	Invaluable cover for many insects and birds; fruits are valuable bird food
Meadow cranesbill (*Geranium pratense*)	Highly attractive to bees
Mints (*Mentha* spp.)	Highly attractive to many insects
Oak (*Quercus* spp.)	Harbours a greater range of wild life than any other British plant, including several hundred species of insect
Rowan (*Sorbus aucuparia*)	Fruits provide valuable bird food; a habitat for many species of insect
Teasel (*Dipsacus fullonum*)	Highly attractive to many butterflies and other insects; seeds provide food for finches
Thistles (*Cirsium* spp.)	Seeds provide food for many finches and other birds
Thyme (*Thymus serpyllum*)	Highly attractive to many insects

* *this is a very personal selection – I have omitted some plants that are valuable as refuges or food for wildlife but really are unsuitable for all except very large gardens (willows and rosebay willow herb for example). The list of flowers attractive to bees and insects is very short but represents those that I recommend most strongly. All wild plants moreover, provide food for at least a few insect species so it may be worth including those that are valuable for any especially rare insects that occur in your area, even if they are less than ideal as garden plants. Local naturalist societies may be able to advise you.*

argument is unresolved; but sycamore with its rapid growth and very large leaves makes a poor tree for most gardens, and I never feel any compunction about removing one, knowing that it has relatively little impact on other forms of wildlife.

But there is another side to the issue of plants harbouring other organisms. The hawthorn bushes that provide a home for raspberry beetles in the early part of the season, and the poplars and willows from which some species of aphid may fly to attack carrots and lettuces, are but two examples. I shall consider all of these issues in more detail in the chapter on garden animal life.

The ecology of fungi and viruses in relation to existing vegetation is also important. Although a few mycologically inclined gardeners may feel disposed to retain certain wild plants in their garden solely because they provide the habitat for some uncommon or interesting species of fungus, they are likely (regrettably) to be in the minority. Of more concern for most gardeners is the role of wild vegetation in harbouring some of the fungi that can cause disease in garden plants. There are two ways in which this can come about. Many of the important disease-causing fungi are fairly conservative in the plants that they attack and only infect a small range of closely related species. But since most garden plants have quite close relatives among native species, there is a good chance that spores of a fungus from the latter will be blown to infect the former. Hence, for example, the removal of brambles from a garden will diminish the impact of some leaf infecting fungi on cultivated blackberries, raspberries and hybrid berries.

In one group of fungi the importance of wild plants can have an additional, more subtle aspect. The fungi known as rusts are often considered to be evolutionarily among the most advanced. Many species of rust have highly complex life cycles, not only involving the production of several different types of spore, but also requiring the presence of two quite different species of plant on which to grow. Part of the life cycle is spent on one type of plant, and part on a second, called the alternate host. The two types of plant are usually quite unrelated but, clearly, eradication of one should protect the second by breaking the fungus' life cycle. The rust disease of irises, which probably has nettles as its alternate host, is one example. Another example with far-reaching effects, although one where two cultivated plant species are involved, is *Cronartium ribicola*, the white pine blister rust, a very serious tree parasite which also infects currants. The widespread growing of currants, especially in areas of Britain where the Weymouth pine (*Pinus strobus*) might be considered for forestry, means that a potentially valuable timber tree has become an unrealistic proposition. The rust disease is also the reason why so few specimen Weymouth pines are to be found in gardens.

It is inappropriate to consider here all the wider aspects of plant disease control, but it is important to appreciate one major principle of plant pathology with ecological implications. Generally speaking, the greater the amount of fungal growth already present on a plant leaf, stem, root or other organ, the greater will be the ability of the fungus to spread and affect other plants. And the sooner the first signs of disease in a garden can be detected, the better will be the chances of containing the outbreak. Once spores spread from the initial focus of infection to establish new foci elsewhere on the plant or on other plants, control, by whatever means, becomes harder in an almost logarithmic progression. The watchword must be vigilance; by detecting fungus early and removing affected plants or leaves by hand (or spraying a chemical in some instances), you can protect the remainder of the garden. There is no better example than the powdery mildew fungus that affects apple trees. This fungus survives the winter on the twigs, and its spores germinate to infect young leaves in Spring. The careful removal of these first infected leaves can be sufficient, on small trees, to diminish the impact of the problem for the whole season.

Before leaving fungi I must refer to one further aspect of their biology that has very considerable, but scarcely appreciated, importance for the ecology of a plant community. Anyone who has afforded mushrooms and toadstools more than a passing glance will realize that certain species are almost invariably found in association with certain species of tree or other flowering plant. This association is more than one of coincidence; it is one of nutrition. Mushrooms and toadstools are spore-bearing bodies, but they are only the outward manifestation of the extensive mycelium inhabiting the soil. The mycelium of some species of fungus forms an intimate association with the roots of certain trees, and a symbiotic relationship, known as a mycorrhiza, develops, in which the fungus provides the medium for the uptake of nutrients from the soil and the passing of them to the roots, while itself deriving some mineral nutrition that it could not otherwise obtain. So important is this association for some plants that they are unable to grow satisfactorily on soil deficient in the appropriate fungal species. Thus, it is possible that some plants may not occur in habitats that otherwise seem appropriate for them, and it may be why some can never be established in certain gardens.

A word should be said about viruses and wild plants in the garden. I have already explained that all viruses are parasitic, and that many cause very serious plant disease problems. Many are catholic in their choice of host plants and are as at home on wild as on cultivated species. The virus known as cucumber mosaic virus is perhaps the extreme example of this and infects thousands of different species. Like many other viruses, it spreads rapidly from one plant to another, often being transferred by aphids or other insects. Notorious wild plant 'reservoirs' of cucumber mosaic virus in gardens are white bryony (*Bryonia dioica*), common mallow (*Malva sylvestris*) and teasel (*Dipsacus fullonum*); but goldfinches may feed on teasels and thus leave cultivated plants alone, and bryony is sometimes recommended as a bee-attractant. I certainly would not advocate bryony, with its poisonous berries as an additional hazard, as a garden plant; but teasel, with its virus and its goldfinches, is but one of many examples to demonstrate that, no matter what some people might tell you, gardening is not simply nature under a modicum of control. There are many occasions when you must make a choice between the cultivated and the wild.

Weeds and weed management

I have referred to the importance of wild plants as harbourers of animal life, viruses and fungi; but of far more direct consequence in most gardens is that wild plants, in the shape of hedgerows, harbour other wild plants, in the shape of weeds. This introduces, therefore, the most important feature of garden ecology, the management and control of weeds.

Weeds, according to the old definition, are simply plants growing in the wrong place – a highly anthropocentric notion for, in one sense, the native weed species have a greater right to call your garden home than do the introduced aliens you have planted. And in fact it was man himself who so largely shaped the environment with his agriculture that he created the ecological niche in which weed species thrive; in many instances, we can only speculate at the habitat occupied by weeds before the bare earth of fields, the mown verges of field edges and the general suppression of the natural dominant species of our ecological formation provided conditions ideal for them.

Before discussing weed management, it is important to appreciate the features that make a successful weed species. Above all, it must be a good competitor for the various ecological niches that constitute your beds, borders and vegetable plots; indeed, it is a feature of many of the most successful weeds that they have evolved with capabilities for growing in a fairly wide range of habitats. Thus, nationally, they build up large populations of individuals able to spread from one habitat to another with few natural barriers. The key to this spread, in most cases, lies with a highly efficient system of reproduction and/or very rapid growth. Many weeds produce large quantities of seeds per plant and, very commonly, these are able to germinate without the need to pass

The rust fungus on willowherbs seems innocuous enough but spends part of its life cycle on fuchsias, one possible reason for treating willowherbs as undesirable in a garden.

The fungus Leccinum scabrum *occurs invariably with birch trees, for its subterranean hyphae form an association (a mycorrhiza) with the tree's roots and aid their uptake of nutrient.*

through a period of winter dormancy. Indeed, some weeds, such as chickweed, (*Stellaria* spp.) and groundsel (*Senecio vulgaris*) germinate and grow so quickly as to pass through several generations in a single season, even though the output of seeds per plant is lower than that of many other species. Even though a period of dormancy may not be needed, many weed seeds can survive for quite long periods, only to germinate when the soil is disturbed, when cultivation practices change (old grassland is dug up, for instance), or when moles, earthworms and other animals bring them to the surface. And, moreover, many species have physiological systems that permit a proportion only of the weed population of the soil to germinate at any given time. There is, indeed, from studies of weed seed survival in the soil, some evidence for the truth of the old adage that 'one year's seeding' leads to 'seven years' weeding' (Diag. p. 111). In such circumstances the garden plant life is swamped, and deprived of light, moisture and nutrient.

Allied to the production of large quantities of viable seed is the possession of an efficient means of dispersing them – by explosive discharge, by wind or water, by man himself in both his small- and large-scale movements, or by animals (either externally, by adhering to their coats, or internally, by surviving the digestive tract).

Nonetheless, not all weeds depend on seed production; indeed, a short list of the most problematic weeds of British gardens (Table 8) reveals that the really serious difficulties arise with deep rooted perennials, with those that produce masses of small perennating structures like bulbils and, of course, with those of them that are little affected by the weedkillers available to gardeners. By its very nature, an invasive underground root or rhizome is highly undesirable in a cultivated plant; so any weed species possessing this method of growth will have an inherent advantage in its competition for the resources of the soil.

Given the criteria for an efficient weed, therefore, what can we learn about weed control or weed management? Although the growth of a population of annual or other freely seeding weeds, or of an individual perennial weed with extensive underground development, is not actually logarithmic, it is certainly an accelerating process as the season (and subsequent seasons) progress. The most important policy for weed management, therefore, is vigilance – 'keeping on top' of the weeds. My maxim for weed control is that a little, often, is very, very much better than a great deal occasionally.

*The lesser celandine (*Ranunculus ficaria*) (above) is one of the most efficient of garden weeds, spreading by means of seeds or bulbils and persisting in the soil with the aid of numerous tiny tubers.*

*Although few 'lower' plants are significant as garden weeds, the field horsetail (*Equisetum arvense*) (right) is one of them. It has deep-seated and very far-creeping rhizomes that are almost impossible to dig up and are unaffected by most weedkillers.*

Those weeds that depend on seed production must be removed before the seeds have been shed, and there are several approaches to this. Purely physical methods such as pulling out by hand, digging out by fork or hoeing, clearly depend on allowing the weed to grow to a recognisable stage and then acting quickly. Contact weedkillers such as paraquat are limited in their real effectiveness to application at the seedling stage. There are, however, a few weedkillers that act by preventing weed seed germination, or by killing the seedling as soon as it appears. One of the most useful chemicals for this type of treatment is propachlor, which is effective for six to eight weeks and is safe to use among most garden plants. As it does not affect weeds once they have emerged from the seed, it should only be applied to weed-free soil. A routine of forking beds approximately every six weeks, and then applying propachlor granules to the freshly dug soil, can render a garden more or less free from annual weeds throughout the season. It is a commentary on the perverted logic of the gardening fraternity, however, that this chemical was withdrawn from sale because the manufacturers couldn't persuade people to use a weedkiller before they could actually see the weeds. Other weedkillers of longer persistence that also kill weed seeds are mostly only suitable for paths and other plant-free areas.

I remained unconvinced that deep-seated or far-creeping perennial weeds such as couch grass (*Agropyron repens*), ground elder (*Aegopodium podagraria*) or bindweed (*Convolvulus* and *Calystegia* spp.) can be effectively controlled without resort to chemical weedkillers. And two products especially have revolutionised this aspect of weed control; glyphosate and, to a lesser extent, alloxydimsodium. Both are systemic or translocated weedkillers; they are taken up by green tissues and moved within the plant's conducting system to effect a kill even of roots and rhizomes many centimetres distant. Glyphosate is a total weedkiller and will kill virtually all types of green plant; it must therefore be applied carefully, although it is inactive in the actual soil. Alloxydim-sodium is especially effective in killing certain grass species and is also valuable as a control for couch.

The degree of specificity of alloxydim-sodium leads me to mention one of the most valuable of all types of garden chemicals, the selective weedkillers for use on lawns. I have described the way in which such common lawn weeds as dandelion and daisy have their meristems sunken in a rosette, and are thus very difficult to combat by purely physical

Table 11   Some of the most important garden weed species, showing their principal methods of reproduction and spread

Species	Life Form	Principal means of Spread
Achillea millefolium (yarrow)	Partial rosette	Rhizomes; some seeds
Aegopodium pogagraria (ground elder)	Earth plant	Rhizomes; some seeds
Aethusa cynapium (fool's parsley)	Annual	Seeds – prolific
Agropyron repens (couch)	Earth plant	Rhizomes; few seeds
Bellis perennis (daisy)	Rosette	Seeds; also short creeping stems
Calystegia sepium (hedge bindweed)	Earth plant	Rhizomes/creeping stems; seeds rare
Capsella bursa-pastoris (shepherd's purse)	Annual/biennial	Seeds – prolific
Cardamine hirsuta (hairy bitter-cress)	Annual	Seeds – explosive discharge
Cerastium holosteoides (mouse ear)	Flopping plant	Seeds – prolific
Chenopodium album (fat hen)	Annual	Seeds – prolific
Cirsium vulgare (creeping thistle)	Bushy plant	Fragile, creeping roots
Convolvulus arvensis (field bindweed)	Earth plant	Creeping roots; some seeds
Epilobium angustifolium (rosebay willow-herb)	Bushy plant	Seeds – prolific
Epilobium montanum (broad-leaved willow-herb)	Bushy plant	Seeds – prolific
Equisetum arvense (field horse-tail)	Earth plant	Rhizomes
Euphorbia helioscopia (sun spurge)	Annual	Seeds – explosive discharge
Fumaria officinalis (fumitory)	Annual	Seeds
Galium aparine (cleavers)	Annual	Seeds – high adhesive
Lamium album (white dead-nettle)	Earth plant	Rhizomes
Lamium purpureum (purple dead-nettle)	Annual	Seeds
Lapsana communis (nipplewort)	Annual	Seeds
Matricaria matricarioides (rayless mayweed)	Annual	Seeds – prolific
Medicago lupulina (black medick)	Annual/creeping	Creeping stems; seeds
Oxalis corymbosa (pink-flowered oxalis)	Earth plant	Bulbils
Oxalis corniculata (yellow-flowered oxalis)	Annual/creeping	Seeds – explosive discharge
Papaver rhoeas (field poppy)	Annual	Seeds – prolific
Plantago spp. (plantains)	Rosette	Seeds – prolific
Poa annua (annual meadow grass)	Annual	Seeds
Polygonum cuspidatum (Japanese knotweed)	Basal bud plant	Rhizomes; some seeds
Ranunculus ficaria (lesser celandine)	Earth plant	Bulbils, seeds, root tubers
Ranunculus repens (creeping buttercup)	Creeping plant	Creeping stems; some seeds
Rumex spp. (docks)	Bushy plants	Seeds – prolific
Sagina spp. (pearlworts)	Partial rosette	Seeds
Senecio vulgaris (groundsel)	Annual	Seeds
Sonchus spp. (sow thistles)	Annual	Seeds – prolific
Stellaria media (chickweed)	Annual	Seeds – prolific
Taraxacum officinale (dandelion)	Rosette	Seeds – prolific
Trifolium dubium	Annual	Seeds; creeping stems
Trifolium repens (white clover)	Creeping	Creeping stems; some seeds
Urtica dioica (stinging nettle)	Bushy plant	Creeping stems; seeds
Urtica urens (annual nettle)	Annual	Seeds – prolific
Veronica filiformis (slender speedwell)	Creeping plant	Creeping stems
Veronica spp. (annual speedwells)	Annual	Seeds – prolific

approaches. However, because the monocotyledonous grasses are botanically unrelated to the dicotyledonous weeds, it has been possible to develop chemicals that work in the opposite way to alloxydim-sodium and kill the latter without harming grass.

Only two representatives from the 'lower' plants are important as garden weeds; horse-tails (*Equisetum* spp.) and mosses. Horse-tails have deeply creeping rhizomes. Because of their unrelatedness to other garden weeds, and also because of their protective, siliceous structure, they are highly resistant to all weedkillers. Many species of moss occur as garden weeds, principally on lawns; they too are unaffected by most conventional weedkillers, but specific moss-killing chemicals are widely available. The growth of moss on lawns should be taken as an indication that the ecological management on the lawn is defective, however. In particular, moss will thrive if the lawnmower is set too low, the grass is shaved and light allowed to penetrate freely to soil level. Additionally, an underfed lawn, a waterlogged and compacted soil, and partial shade may mean that the grass is unable to compete effectively with the moss.

Having returned to more basic ecological principles, I shall now describe two features of weed management that will be especially appealing to those who eschew the use of weedkillers, and which provide a useful adjunct to the hoe and the fork. The first is to deny weeds the light, air and moisture that they need to thrive. While a deep-seated perennial with an established underground growth will always have enough vigour to reach the soil surface, annual weeds can be very effectively smothered with a thick blanket of mulch laid among established herbaceous perennial and shrub borders. Peat, well-rotted manure or compost, or shredded bark are all suitable for the purpose. Secondly, although I have explained that weeds *are* weeds because they are such effective competitors with garden plants, it may be possible to choose some cultivated species and varieties that can at least 'hold their own'. Working on the principle that most annual weeds have a high demand for light, garden plants with large leaves, and that grow rapidly and therefore soon result in a closed canopy with no bare soil, can sometimes suppress them. The use of a potato crop to 'clean the ground' is an old gardening practice, although some cultivars are more effective than others. In the herbaceous border, dahlias are very effective suppressors of weeds, but only from late July onwards. Thus a management system is needed that relies on the hoe or a chemical in the earlier part of the season until the leaf canopy is closed. Paeonies, on the other hand, have very long period in leaf and compensate for a fairly short flowering season with about eight months of reliable suppression of annual weeds.

Conversely, there are some garden plants that are particularly poor competitors; onions, especially, germinate and grow very slowly in the early stages and have a narrow, upright form with consequently poor ground cover. Onions also, incidentally, have very shallow root systems and are among those plants between which the hoe should not be used.

There is, therefore, no overall answer to the subject of weed control, but an understanding of the ecological requirements of both weed and garden plant should help in the devising of a management programme, incorporating physical methods and/or an appropriate weedkiller.

The introduction of wild plants into the garden

Finally, then, to the vogue for so-called wild gardens; sites where the natural vegetation is not merely incorporated into the garden (or removed because it competes with it), but actually serves as the chosen plant life. I shall not consider what I prefer to call the wilderness, the site that may once have been a garden but which has been allowed to be colonised *ad lib* by native plants and so resume its progression towards some distant ecological climax. Anyone can create a wilderness by the simple expedient of neglect; but it is not a garden. In my definition, the wild garden is a blend of native and cultivated plants, chosen for their beauty, for their role as micro-habitats for other wildlife, or because of some deeply felt conviction that this is the way to preserve rare plants from extinction. The latter may, in some cases, be true, but it falls as far short of performing the ideal role as zoos do with animals.

It will, I hope, be apparent from all that has gone before in the book regarding the generally unbalanced competition between 'weeds' and 'cultivated plants' (and remember that, in a wild garden the 'cultivated' plants are the 'wild' ones) that, far from being an easy option, the wild garden is intrinsically one of the hardest of all to manage. Advice regarding the establishment of wild flower gardens seems to have become a minor industry in recent years, and I shall confine myself, therefore, to some personal comments on the subject. First, if you encourage those species that possess the attributes that I gave for the successful weed, you will create problems in plenty. Second, I believe that wild plants should take second place in quantity and planted area to cultivated plants. Third, it is virtually impossible, in most gardens, to establish a wild plant garden without first being prepared to use some weedkiller.

While wild plants can be used within a herbaceous border or shrubbery, in exactly the same way as cultivated plants, it is far more satisfying and instructive to designate a defined area for wild plants alone. The size of this will depend on your total available space and inclinations, but it does not really matter what type of habitat it is; wild species can be selected appropriate to grassland, woodland or even wetland. If the chosen area has experienced little cultivation, there is some merit in leaving nature to take its course, and see what wild species grow. It may be you will find the basis of an interesting flora that can be regulated by careful removal of some species and addition of others. Clearly, if you adopt this course with an old vegetable plot you will achieve little more than the results of the germination of a large population of weed seeds. In such cases there are basically two ways to proceed: either sow the area with a commercially prepared mixture of wild plant seeds appropriate to your chosen habitat (many seed companies sell 'meadow mixtures', 'woodland mixtures', etc.), or raise plants individually from seed and plant up the area piecemeal. A third option is to sow with a mixture of wild grasses and then plant individually raised plants within this sward.

An old vegetable plot is actually the worst possible site for wild plants. Earlier in the book, I discussed the abnormal growth demands that we place upon our specially selected and bred vegetables, demands that can only be met by intensive fertiliser usage. Such well-manured or fertilised soil is not the environment of wild plants and will result in some species, especially some of the more vigorous grasses, luxuriating at the expense of others. If you have a choice, therefore, select the part of the garden in which the soil has experienced the greatest neglect; and, whatever else you do, use no artificial fertiliser.

Above all, before embarking on a wild plant garden, look, and look carefully, at a natural habitat (or what passes for one in Britain today). See the way in which individual plants and individual species contribute to the harmony of the whole; but don't imagine for one moment that you can freeze the ecological evolution of this environment once you have attempted to recreate it in your garden. A woodland or an old meadow are just as much part of an ecological succession as the herbaceous border or the formal bedding. The implication, so often given, that by mimicking nature you will not only create something inherently better than a garden of cultivated plants but also something of greater durability, is an ill-informed folly.

Chapter Six

Cultivation and cultivated plants

Mankind has had a long and intimate relationship with the plant kingdom; plants have served us and our ancestors well as food, shelter, fuel and as the raw material for a large proportion of our artefacts. The artificial management of plant growth has been a feature of human society for at least 10,000 years – ever since someone realised that the more or less random gathering of wild plants and plant products was considerably less efficient than deliberately sowing or planting them close to the village. This seems to have occurred first in the region presently occupied by Israel, Iran and Iraq, and spread later to other parts of the eastern Mediterranean. Later still, the organised growing of crops began in India, in China and in South America so that by around 4,000 years ago it was an established feature of societies in many parts of the world. This early growing of crops was, according to our modern concepts and definitions, farming rather than gardening, although there is no hard and fast division between the two. Nonetheless, today, a garden tends to be thought of as an area adjoining a dwelling where plants are grown solely for the needs of the inhabitant, rather than for sale or trade. It has also come to be thought of as an area at least as much for ornament as for the provision of food, and precisely when the use of gardens for decoration began seems obscure. Quite possibly it arose through the careful choice of some of the more attractive among edible crops, herbs for instance, or, I have often thought, through the deliberate growing of appealing wild species that arose as weeds among vegetables and cereals.

I shall describe the ecology of the many different types of garden in Chapter 8, but I have introduced the word 'choice' and this, together with 'selection' is the key to the differences between the collecting and gathering mentality and that of the cultivator. For the cultivator imposes his will on nature by causing plants to grow where naturally they did not, by choosing plants to suit his needs and desires rather than those that nature dictates, and by placing in juxtaposition species that naturally may never occur together. The cultivator chooses, moreover, not only the space within which plants will grow, but also the time that they will be planted and harvested. Early on, he learned two things. First, that some plants were more amenable than others to having their natural inclinations disrupted and, second, that almost all plants whose natural regime was disturbed would need compensating in some way for their upheaval. To germinate at the required time, to grow satisfactorily

A fairly high proportion of ornamental trees are grown as species, unaltered by breeding and selection from those occurring in the wild. The beautiful early flowering Oriental cherry Prunus sargentii *is one of them.*

where other crops had grown the previous year, to reach maturity and ripeness at the desired time, all necessitated artificial watering, manuring and general cossetting to the extent of suppressing those plants, the weeds, that grew in company with the chosen few. Man soon learned some of the lessons in basic plant ecology that I have discussed in the preceding chapters.

At some rather imprecisely determined date, however, the elements of choice and selection became manifest in a very important manner beyond space and time. They became manifest in kind too, for the more observant among the early agriculturists and gardeners began to see the variation that was to excite Charles Darwin so much many generations later. Not only were there differences between species, there was considerable variation between the individuals of species too, which showed itself as uneven maturing times among their vegetables, uneven sizes of fruits and flowers, perhaps even the differential susceptibility of different plants to the pests and diseases that were the accompaniment to cultivation from the beginning. When the observant farmer or gardener displayed intelligence, intuition, or merely curiosity, and began to select the plants with more desirable features and save seed from them alone for the next season's sowing, we really had the beginning of something hugely important. Here was one organism, man, altering the speed of evolutionary change in another organism, the plant, in a concerted and directed manner. Selection that could have taken generations by normal processes of competition, success and failure, suddenly became immensely fast. Now, after many thousands of such deliberate selections, most gardeners can look at most gardens and see an array of plants that are almost all the results of some form of human interference in their development. Only in the garden of the serious plantsman or the botanist are you likely to see more than a handful of unaltered, or natural species, appearing in the same form as they do in their natural habitat.

Before considering the ecological and botanical consequences of this turn of events, it is important to look at the next, logical step to arise after the process of artificial selection. Again at some unknown historical date, someone, somehow, discovered the technique of hybridisation. It is possible that some unusually gifted and observant individual (or, indeed, individuals, for the process was clearly discovered more than once) noticed that where two different types of plant were growing closely together, seed collected from them produced some seedlings intermediate in character between the two. From then on, it was a matter of understanding why bees and other insects visit flowers, of appreciating that two separate flowers or even two separate plants are often needed before seed can be produced at all, of deliberately manipulating the pollination process to ensure that it occurred only between chosen parents and, *eureka*, the science of artificial plant breeding had arrived. Indeed, if my typical gardener looks again at his typical garden, not only will most of the plants he sees be different from wild species, they will be hybrids too. Modern gardening consists largely of the cultivation of hybrids, and thus it is essential to look in more detail at the biological advantages and disadvantages of hybridism and to set them in an ecological context. The way is somewhat obstructed by nomenclatural hurdles, however.

My dictionary defines a hybrid as 'the off-spring of two animals or plants of different species or varieties'. Already we seem to be in murky waters, for what (we may wonder) is the difference between a species and a variety? And what, come to think of it, is a sub-species, a genus, a cultivar, or even a clone?

The species is the cornerstone, buttress (or other appropriately robust metaphor) of the whole of genetics, plant breeding, evolution and, in large measure, of biology itself. A species is, nonetheless, not amenable to precise definition, although it can conveniently be thought of as a group of individuals readily able to breed among themselves but not generally able to breed with organisms in other groups or species. A species, therefore, is a grouping of organisms reproductively isolated from other organisms. Sometimes, further reproductive isolation (usually through geographical separation) of some of the individuals of a species may result through evolutionary change in a group with some slightly different characteristics from the majority, but still able to interbreed with them should the opportunity arise. Should this opportunity be denied for a long time, however, the isolated small group, or subspecies, may well lose this interfertility with its progenitors and thus become a new species in its own right. But it is impossible to lay down rigid rules of nomenclature, for such a system, based as it is on relative interfertility, cannot always be put to the test: groups of organisms geographically isolated, and morphologically dissimilar, *may* be able to interbreed with each other if they are given the chance. Indeed, the hybrid itself is living proof that some individuals, of some species, can interbreed with those of other species, although the fact remains that, under natural conditions, normally they do not do so. Groups of species that display similarity in many ways are linked in classificatory schemes into a genus (plural, genera; but please note that the word species is the same in both singular and plural).

The difference between the variety and the cultivar is obscured in most gardening literature to the point of being cryptic, but is most conveniently remembered if the word cultivar is expanded as cultivated variety. Thus, a natural variety is a population within a species or a sub-species that differs in some distinct way from other populations but is still interfertile with them and does not differ from them to the same degree as does a sub-species from a species. A cultivar is just such a group of individuals that has arisen either accidentally or through artificial hybridisation and selection during cultivation. Table 12 gives examples of genera, species, sub-species, varieties and cultivars of some common garden plants.

Before leaving this nomenclatural tortuosity, I must return briefly to the clone, for this is a very important category in horticulture. It is a group of individuals derived by vegetative or non-sexual reproduction from a single individual; they are all alike genetically and display not even the variation that occurs within the individuals of a species. The plants themselves may differ by virtue of being grown in different soils or in different ways, but genetically they are identical. The category is important, for many of the most significant garden plants are not routinely perpetuated through the medium of sexual reproduction and seed. The many different garden 'varieties' of roses, potatoes, dahlias and carnations are all examples of plants that exist in the form of clones. They possess the attribute of genetic uniformity therefore, but this is a two-edged sword, for while this uniformity may guarantee similarity of flowers and fruit, it also guarantees similarity of susceptibility to disease. Vegetative reproduction also guarantees the passing on, from one individual to the next, of any virus contamination present in its tissues. Hence the need for regular replacement of potato stocks, for example, with fresh, virus-free tubers.

And so, to return to hybrids and their special attributes. I shall not consider the many genetic complexities that govern how particular characters are inherited from one generation by the next, and of the likelihood of chosen features being manifest in the next and/or succeed-

Table 12 Examples of the way that cultivated plants are named

Family: always ends in -ae. Rosaceae; Orchidaceae; Compositae; Gramineae

Genus (pl. genera:) always begin with a capital letter. *Rosa*; *Iris*; *Lactuca*; *Malus*

Species: always follow a genus name (which may be abbreviated to its initial) and begin with a small letter. *Rosa moyesii*; *Iris histrioides*; *R. ponticum*; *P. sativum*

Sub-species and Variety: always follow the species name. *Brassica oleracea capitata*; *Cedrus atlantica glauca*; *Spiraea japonica fortunei*; *Lactuca sativa crispa*

Cultivar (including clone): usually follow the species and/or variety name but may be used with the genus name alone, especially where the cultivar is a hybrid. *Rhododendron* 'Pink Pearl'; *Brassica oleracea gemmifera* 'Peer Gynt'; *Magnolia grandiflora* 'Goliath'; *Salix alba* 'Vitellina'.

Hybrid: simple hybrids between two species are usually given a name prefixed by 'x', which may then be followed by the cultivar name. *Acer × dieckii*; *Cistus × florentinus*; *Laburnum × wateri* 'Vossii'; *Syringa × chinensis* 'Alba'.

The modern garden petunia (right) like most bedding annuals has changed through many years of breeding and selection, particularly for colour variation, until very different from its wild ancestor.

Abutilon striatum *var.* thompsonii *(above) is unusual in being perpetuated vegetatively solely to maintain the virus contamination that gives rise to the attractive leaf mottling.*

ing generations (although they should not be minimised, for plant breeding is a difficult and frustrating discipline). Rather, think first of hybridisation as a means of combining some of the features of two separate plants in one individual. Two similar species may, perhaps, include, in the one, some individuals with large flowers and, in the other, some with many flowers. Artificially cross the two, weed out the individuals that don't match up to your criteria, collect seed from those that do, and, in time, you should arrive at a population with flowers both many and large. Provided these characteristics remain relatively constant in your new population when its individuals are allowed to cross freely among themselves, you can congratulate yourself on having produced a new cultivar, and name it accordingly.

Having made your crossings and selections for the features that appeal to you, you will quite probably have unwittingly eliminated some other feature that enabled the species to flourish in competition with other plants, and the lack of which they may now find a disadvantage. Larger flowers, and especially double flowers, could well mean that some of the floral parts intended to aid seed production have been modified to the extent of being visually colourful and appealing but functionally useless. Your cultivar may produce fewer seeds, therefore, and present you with a continuing difficulty in perpetuating it. Although you have used the technique of hybridisation in arriving at your new cultivar, the cultivar itself is maintained in a pure form by what is termed open pollination; a group of individuals is simply allowed to grow together and pollinate each other.

What I have said applies strictly to plants known as cross-pollinated types, those where pollen from one plant must be transferred to flowers of another to effect fertilisation. Some species or varieties, however, are self-pollinated; only one individual is needed and thus the plant breeder produces his cultivar by bulking together the seed produced by many individual plants, carefully chosen for their similarity.

The individuals of cultivars produced in the ways described are still very variable, even though certain desirable features have been incorporated into them. In time, commercial plant growers, especially of vegetables, began to demand from their scientific colleagues plants with greater uniformity, both in yield and in maturing time. In essence, it became increasingly inefficient and costly to cut a cabbage crop, for instance, in several visits, spread over several weeks, as fresh batches of plants reached maturity. This trend became even more marked with the introduction of automated harvesting systems which required entire fields of plants to be at identical stages of development. The result of this demand, over the past few decades, has been the F_1, or first generation hybrid. The method of producing an F_1 hybrid cultivar is compared with that of producing an open pollinated one in Diag. p. 155 and it must be stressed that the hybrid production is highly labour intensive, and hence the seeds are more costly. The pollination process must be performed anew for each fresh batch of seeds, moreover, because the seeds produced by an F^1 hybrid individual if left to itself would be a genetic hotch-potch and quite useless as garden plants.

Not all types of vegetables, let alone ornamentals, are available as F_1 hybrids, either because of genetic difficulties in their production or because the open pollinated types already display acceptable uniformity. Hybrids are exceedingly difficult to produce in markedly self-pollinating plants such as lettuces and French beans, and it is evident that the returns would probably not repay the efforts involved in trying to develop them.

Old cultivars giving rise to very variable fruits

Best plants allowed to mass mollinate together and their progeny then selected and tested

Best plants deliberately inbred and their progeny then selected

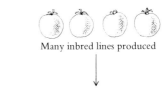

Many inbred lines produced

The best offspring allowed to mass pollinate together

Inbred lines crossed together in all combinations

After many selections, a new open-pollinated cultivar

Progeny tested, and the best of them give new hybrid cultivars

The differences in the processes used to produce new open-pollinated cultivars (left) and new F$_1$ hybrid cultivars (right). The hybrid production is highly time consuming and labour intensive and the crosses between the inbred lines must be performed afresh each season. Hence the resulting seeds are usually much more costly than are those of open-pollinated types.

Generally, the hybrid plant, as with the cross-breed or mongrel animal, displays good competitiveness with the pure species or the open-pollinated cultivar, a phenomenon given the name hybrid vigour. Although vigour is, in itself, a fairly meaningless term, it is useful in conjuring up the image of a plant well able to take care of itself in the hurly-burly of the ecosystem, although the precise manner in which this is manifest could range from a deeper reaching root system or a greater leaf area, to a more robust stem better able to withstand wind damage.

I referred earlier, when describing clones, to the matter of uniformity, and this is equally applicable to F$_1$ hybrid cultivars. Certainly, they possess the uniformity of size and maturity so beloved of the commercial grower, and this, combined with hybrid vigour very often gives them excellent ecological competitiveness; the more efficient weed suppression, for instance, of some hybrid cultivars is very marked. Again, as with clones, however, uniformity can be a mixed blessing, for if uniform maturity and size is combined with uniform susceptibility to a particular pest or disease there can be devastating consequences. Even uniform *resistance* to such problems is often of little merit, for the pest or disease species, with a genetic constitution broader than that of the highly selected crop, is much better equipped to find within its population, or to produce by mutation, individuals able to overcome this resistance. If a monoculture is an invitation to ecological problems, as I outlined in Chapter 1, a monoculture of an F_1 hybrid can be an open door to a disaster. It is part of the spiralling inevitability of modern plant growing that the consumers' and growers' desire for plants as alike as the proverbial peas in the pod, has resulted in an ever increasing demand for the pesticides needed to protect those plants from enemies that simply pale into insignificance in the mixed ecosystems of the wild world.

Do my comments mean that gardeners should grow F_1 hybrids for their generally good competitiveness with weeds, larger yields and more showy flowers, or avoid them for their corresponding disadvantages? There is no overall answer, but it should be remembered that hybrid cultivars were produced originally because of the dictates of commercial growing, and that having all your Brussels sprouts maturing at the same time might not be as desirable in a garden as it is in a commercial field.

One disappointing aspect of modern plant breeding and production that is, admittedly, not entirely to be laid at the feet of F_1 hybrids, is the general fall in the diversity of cultivars offered for sale by seed companies and nurserymen. The maintaining of plant cultivars in pure form and marketable quantities is an expensive undertaking. It must be conceded that many of the older forms make much poorer garden plants than their modern counterparts; nonetheless, I have already referred to the fact that in breeding and selecting for certain desirable characters, we almost certainly eliminate others. And who is to know what value should be placed on those characters? Suppose, for instance, that a particular plant species is selectively bred for larger and more shapely flowers, and that those individuals selected for propagation have slightly less hairy leaves than the plants with poor blooms. Who will take note of this, or deem it important in any way? Of course, no-one will, and so, with the passing of time, the hairy old types disappear at the expense of their more floriferous and glabrous companions. Then, one day, in the course of investigating why the newer cultivars of our plant have suddenly become very prone to virus problems, it is realised that the hairless leaves of the new plants are very much more attractive

to aphids for their feeding activities. The aphids carry the virus and so the cry goes up, 'Find me some plants that are hairy, so that the aphid/virus resistance may be bred back in again'. But the cry comes too late; the old types have gone forever and taken their irreplaceable genes with them. Who today can show me seeds of cucumber 'Hero of Surrey', 'Brown Silesian' lettuce or 'Ellison's Gigantic' broccoli, all picked at random from a catalogue of 1849? Barely in time has the significance of such things come to be appreciated, and so-called gene banks are being established widely where old cultivars, of no apparent value other than as curios, can be maintained in perpetuity as investments for some future need. Already many old cultivars have gone and, indeed, even now, in less commercially important species (ornamentals especially), there is little effort being made at genetic conservation.

It is, I hope, becoming apparent that I believe diversity, taken in its widest sense, to be a gardening virtue based on sound ecological principles. Diversify the genetic base of your plants as far as possible by growing at least *some* non-hybrid cultivars; diversify by growing a range of cultivars of each type of plant, and by growing many different types of plant; and diversify the demands you make on your soil, by crop rotation.

It is now appropriate to consider more closely the various ways in which garden plants are grown together, trained and managed, and see how each manipulation makes use of or opposes the plants' natural inclinations.

Beds and rows

Apart from the predominance of a restricted number of types of plant, the feature of many gardens (especially vegetable and fruit gardens and formally planted flower beds) that most obviously sets them apart from natural plant associations is that the plants are arranged in more or less straight lines. Other than along a straight river bank or the edge of a cliff, it is most improbable that you will see plants growing naturally in this way. With the summer flowering annuals and biennials generally called bedding plants, the reason is usually to create the particular visual effect that some people clearly find pleasing. With vegetables, the reason is different and largely historical.

Before the advent of mechanical aids, farmers sowed their seed by hand; they walked across the field scattering or broadcasting it as they did so. In some parts of the world, this technique is still followed; but it is biologically inefficient. It is impossible to obtain a uniform distribution of seed in this way; some falls sparsely, with large gaps between the emerging plants, some falls into clumps where the young seedlings must compete with each other for available resources, some falls on good soil, and some, as the Bible reminds us, falls on stony ground. Nonetheless, this system prevailed for several thousand years until a Berkshire barrister turned farmer called Jethro Tull turned over tradition with his plough and, in the early eighteenth century, invented a horse-drawn drill that placed seed in equidistant rows. He also devised a horse-drawn hoe that was worked between the rows of crops to destroy weeds. Tull demonstrated that his crops, grown in rows, gave better yields than those raised from broadcast seed. The technique passed from corn to vegetables and so to gardens where, although few of us employ horses, the straight rows have persisted.

Not until fairly recently did anyone question Tull's deduction that plants gave higher yields *because* they were grown in rows. Had Tull been an ecologist, he would have known better! Of course his plants in rows yielded more; but not because of any intrinsic merit in the row itself, simply because his horse-drawn hoe was removing the competing weeds. In a broadcast crop, all weeding had perforce to be done, very much less efficiently, by hand.

Examine a typical, traditional seed packet and you will read directions for the spacing required between the rows and for the within-row distances between individual seeds (or between individual plants after the seedlings have been thinned). Then take a good, long look at a typical traditional vegetable garden and see what this means in practice. The plants within the rows are much closer together than are the rows themselves, with the result that, whereas few weeds appear in the rows, between them is a no-man's land where weed growth is rampant. These weeds must, moreover, be controlled, but to reach them all, even with a long hoe, you will need to walk along some of the inter-row paths. The result, by the end of the season, could well be a soil compacted to the condition of concrete and effectively impeding the penetration of water, air and fertiliser. To grow vegetables in the traditional manner, in rows, is to fly in the face of almost every basic principle of plant ecology. Accept, as we must, that the vegetable garden is far removed from the natural plant association, but don't abandon your ecology to boot. Give yourself and your plants a better than even chance against

Jethro Tull's corn drill
Tull, with his horse-drawn corn drill and inter-row cultivator revolutionised agriculture in the early eighteenth century. He bequeathed an obsession with growing plants in rows that has persisted until very recently, despite its ecological inefficiency.

the weeds and ensure that the soil doesn't form a compacted crust – by adopting the deep bed system of growing. If vegetables are grown in beds approximately 1.2 metres wide, with pathways between, you will find that the whole of the bed can be reached for weeding without the need to walk on it at all. Thus, the soil doesn't become compacted, and thus there is no need to dig it every year. Nonetheless, it is no use earmarking any random patch of garden, bringing a halt to annual digging, and calling it a vegetable bed. The bed must first be rendered 'deep'. Initially, therefore, it must be dug to two spades' depth and plenty of well-rotted farmyard manure or compost incorporated. Thereafter, the site can be topped up with organic matter each autumn and it should not be necessary to dig it again for at least four or five years. If the 1.2 metre width presents something of a stretch, place bricks either side, lay a plank across, bridge fashion, and work from this. Within the beds, it is perfectly possible, and often most efficient, to space the plants equidistantly, but some recommendations for vegetable spacings are given in Table 13.

Trenches

Trenching has two rather different meanings in gardening. The trench traditionally used for such vegetable crops as runner beans and for a few flowers, such as sweet peas, is really a means of providing good soil structure, moisture retentiveness and nutrient supply for plants that are naturally deep rooted. It is used essentially for plants grown in rows (and here is one instance where row cropping does provide the most efficient use of the land) and is normally prepared by digging to two spades depth in the autumn, before a spring planting or sowing, and incorporating manure or compost and fertiliser within it. The principle is that of a localised and deep soil amendment, such as I described in Chapter 3, but it is most important to remember that the organic matter and fertiliser must be incorporated thoroughly throughout the trench. Some gardening books still advocate layering the manure in the bottom of the trench alone and this is not only wasteful of resources, but can be counter-productive in that it may encourage a water-logged region to develop, which may later give rise to root-rotting conditions.

Superficially, the most illogical of all planting operations is that employed in the second type of trenching, used typically for blanching celery. Here, a trench is dug, but the plants are placed in the bottom of it, and increasingly buried as they develop. No plant grows naturally in

Only people plant in straight lines. . . . In this instance, a previous owner's desire for a windbreak has had a ludicrous outcome.

Table 13	Optimum spacings for some vegetable crops	
Crop	**Spacing (cm)** Between row	Between plants
Beetroot	18	10
Broad Beans	23	23
Dwarf Beans	23	23
Runner Beans	60	15 (single rows)
	30	15 (double rows with 45 between each pair)
Brussels Sprouts	75	75
Spring cabbage	30	30
Summer cabbage	45	45
Winter cabbage/Savoy	45	45
Calabrese	45	45
Carrots	15	10
Early Cauliflower	55	55
Autumn/Winter Cauliflower	90	90
Celery (self-blanching)	25	25
Leeks	25	15
Lettuce	25	25
Bulb Onions	25	15
Salad Onions	15	3
Parsnips	30	15
Peas	12	12 (in triple rows with 45 between each trio)
Potatoes (medium sized tubers)	90	25–40 (depending on cultivar)
Bush Tomatoes	48	48

this way, of course, and it serves primarily to deny the leaf stalks of light. In such conditions the green colouring matter chlorophyll fails to form, the tissues are more or less white, and the plant, incidentally, is inclined to be more tender and tasty. Most gardeners find celery grown in this way preferable to the so-called self-blanching types, which are planted more conventionally on the surface and the inner leaf stalks denied light simply by virtue of growing closely compacted together.

The principle of blanching crops by denying them light to render them more palatable is applied in varying ways to other vegetables such as sea-kale, dandelions, cardoons, some types of chicory, and rhubarb. It was more popular in times past when such forced and blanched crops provided almost the only source of fresh vegetables in the winter. Modern methods of preservation like freezing and canning have relegated most of them to the category of curiosities.

Orchards

I use the term orchard to describe a collection of fruit trees, although many of the basic principles of orchard management are applicable to single trees also. Where trees are trained against a wall, or are other than free-standing individuals, however, the comments given on p. 178 regarding wall shrubs are relevant too.

The purpose of an orchard is to produce fruit, and whether the trees are apples, pears, plums or cherries, their blossoms require pollinating. Failure of pollination can be due to several reasons, of which the commonest ecological cause is that the trees are in too windy a situation, with the result that pollen and/or pollinating insects are blown away before they have had a chance to achieve their purpose. Shelter is of special importance around fruit trees, therefore.

The second cause of a failure of pollination is that no compatible pollen is available. Most cultivars (or, more strictly, clones) of apples, pears and plums are not self-fertile. Generations of artificial selection and hybridisation have resulted in this unnatural genetic state, and it may be overcome by growing clones in specific combinations; Table 14 gives some suitable combinations for many of the most commonly grown types of the major tree fruits.

Fruit trees, like other plants, are also subject to disease; whereas the consequences of disease in an annual or relatively short-lived or herbaceous perennial are of limited impact, a diseased tree is a long-term problem. The most important of the diseases, moreover, are those that like the trees themselves, are perennial. The futility of planting healthy new trees in proximity to old and cankered individuals must be stressed, therefore, for the disease will most assuredly spread from one to the other. And, for reasons that I mentioned earlier, plants propagated from virus-contaminated parent stock will themselves carry this contamination; only young trees raised from certified virus-free stock should be used.

The repeated selection of fruit trees for their fruiting qualities has had a further consequence in that many clones produce root systems incapable of supporting the plants. Because of this, most fruit trees are now grafted onto root-stocks of quite different clones, separately and specifically bred for their rooting qualities. Choice of rootstock is as important as choice of fruiting type, for the rootstock confers on the plant the overall potential size of the tree. Table 15 lists the range of rootstocks available for the major tree fruits, with their cultural characteristics.

There is one further major feature of orchard ecology to consider. I have mentioned earlier in the book two consequences of growing fruit trees in grass: one is that grass beneath trees limits heat loss but may render blossom prone to spring frost damage; the second is that grass cover minimises the likelihood of soil contaminated with disease organisms being splashed by rain onto low-hanging fruit. There is a third factor to consider, however: grass is a very effective competitor with the tree for nutrient and water. A young tree grown with grass to the base of its trunk may be functioning very much less efficiently than one grown in bare soil. Conversely, grassing up to the trunk of a more mature and over-vigorous tree can be effective in slowing down its growth. On balance, plums and also apple trees on dwarfing rootstocks are best planted in a circle of bare soil, about 1 m in diameter, and not grassed closely for about three years, apple trees on more vigorous rootstocks not until about five years, while pears benefit by not being grassed down until well mature.

Apple canker disease
This very damaging and difficult to control problem is one very sound reason for not planting new trees in old orchards unless the existing stock is really disease-free.

Soft fruit gardens

So-called soft fruits belong to two distinct plant families, the Rosaceae and the Grossulariaceae and display a wide range of growth habits and ecology. Nontheless, they are normally grown in close proximity for reasons of convenience and because all benefit markedly from effective protection from birds. This is best achieved by erecting a purpose-built fruit cage, with wire-netting sides and with nylon or other lightweight netting top. There are reasons, besides weight, for not using wire as a cover. Damage to plants can result from toxic zinc residues dripping onto them from galvanized wire; and it is convenient to be able to remove the cover in regions where heavy snow falls are to be expected, and where difficulty may be experienced in obtaining satisfactory pollination. Sometimes, pollinating insects are reluctant to pass through the wire sides of a fruit cage and, in such an eventuality, removal of the cover for a few weeks during flowering will be beneficial.

Most types of soft fruit are fairly shallowly rooted and any disturbance of the soil around the plants will be harmful. Weeding should be by very light surface hoeing or by the careful use of weedkillers, therefore, rather than by digging. Because of the surface rooting habit, moreover, most types of soft fruit benefit from mulching early in the season to aid moisture retentiveness, and from supplementary watering as the fruit swell.

Table 14   Some suggested combinations of fruit tree clones to obtain good pollination and fruit set

Fruit	Recommended pollinator	Fruit	Recommended pollinator
Dessert Apples		**Pears**	
'Beauty of Bath'	'Greensleeves' or 'Idared'	'Conference'	'Onward' or 'Louise Bonne de Jersey' ('Conference' will set some fruit on its own but it always better with a pollinator)
'Blenheim Orange'	'Greensleeves' or 'Discovery'		
'Crispin'	'Cox's Orange Pippin' or 'Discovery'	'Doyenné' du Comice	'Conference' or 'Onward'
'Cox's Orange Pippin'	'Discovery', 'Greensleeves' or 'James Grieve'	'Louise Bonne de Jersey'	'Conference'
'Discovery'	'Cox's Orange Pippin' or 'Greensleeves'	'Onward'	'Conference' or 'Williams' Bon Chrétien'
'Ellison's Orange'	'Cox's Orange Pippin' or 'Greensleeves'	'Williams' Bon Chrétien'	'Conference' or 'Onward'
'Fortune'	'Greensleeves', 'Discovery' or 'James Grieve'	**Plums**	
'Greensleeves'	'Discovery' or 'Grenadier'	'Cambridge Gage'	'Czar' or 'Marjorie's Seedling'
'Idared'	'Cox's Orange Pippin' or 'Discovery'	'Czar'	Self-fertile
'James Grieve'	'Cox's Orange Pippin' or 'Discovery'	'Marjorie's Seedling'	Self-fertile
'Jupiter'	'Discovery', 'Spartan' or 'Sunset'	'Rivers Early Prolific'	'Czar' or 'Victoria'
'Kent'	'Cox's Orange Pippin' or 'James Grieve'	'Victoria'	Self-fertile
'Laxton Superb'	'Cox's Orange Pippin' or 'Greensleeves'	**Damsons**	Self-fertile
'Redsleeves'	'Greensleeves' or 'Fortune'	**Cherries**	
'Spartan'	'Greensleeves' or 'Discovery'	'Morello' (cooking)	Self-fertile
'Sunset'	'Cox's Orange Pippin' or 'James Grieve'	'Stella' (sweet)	Self-fertile
'Tydeman's Late Orange'	'Greensleeves' or 'Spartan'	**Apricots**	
'Worcester Pearmain'	'Cox's Orange Pippin' or 'Greensleeves'	'Moorpark'	Self-fertile but use a paint brush to aid set
Cooking Apples		**Nectarines**	
'Bramley's Seedling'	'Spartan' plus 'Discovery'	'Lord Napier'	Self-fertile but use a paint brush to aid set
'Grenadier'	'Discovery' or 'Greensleeves'	**Peaches**	
'Lord Derby'	'Spartan' or 'Tydeman's Late Orange'	'Peregrine'	Self-fertile but use a paint brush to aid set
'Rev. W. Wilks'	'Idared'	**Medlars**	
		'Nottingham'	Self-fertile
		Quinces	
		'Vranja'	Self-fertile

Table 15   The major fruit tree rootstocks and their characteristics

Fruit	Rootstock	Characteristics
Apple	M. 27	Very dwarfing [tree 1.5 m]; only for very good soils; requires permanent stake
	M. 9	Dwarfing [tree 2 m]; best for general garden use; requires permanent stake
	M. 26	Semi dwarfing [tree 2.5 m]; good for poorer soils
	MM. 106	Semi vigorous [tree 3–4 m]; better than M. 26 for weaker clones
Pear	Quince C	Moderately dwarfing [tree 3–5 m]; best for general garden use but not for poor soils
	Quince A	Moderately vigorous [tree 4–5 m]; best for poorer soils
Plum	St Julien A	Semi dwarfing [tree 3 m (by pruning)]
	Pixie	Dwarfing [tree 2–3 m (by pruning)]
Cherry	F 12/1	Vigorous [tree to 10 m]
	Colt	Semi vigorous [tree 3 m (by pruning)]
Damson	St Julien A	See plum
	Pixie	See plum
Peach	St Julien A	See plum
Quince	Own roots	
Medlar	Quince	

M.27   M.9   M.26   MM.106

Relative growth rates of the same apple cultivar after twelve years on different rootstocks – the tree on MM.106 is 3.5 m tall

Bush and cane fruits produce their fruit principally in one of two ways; either from flowers produced on shoots that grew during the previous year, or from flowers on the current year's shoots; just as with roses and many other types of woody garden plant, a knowledge of the method of flower or fruit production is essential in order to decide how and when the plant should be pruned. I shall consider the subject of pruning on p. 183, but it should be appreciated that there may be little consistency in flowering within plant genera, and soft fruit illustrate this particularly well. Raspberries, like blackberries, belong to the genus *Rubus*, and the fact that the species in the genus are closely related is evidenced by the numerous hybrids – loganberries, phenomenal berries, boysenberries and tayberries to name a few. Modern raspberry varieties themselves are inter-specific hybrids, most of the summer fruiting cultivars having the native *R. idaeus* and several exotic species in their parentage. These all produce their flowers and fruit on the previous year's wood; but some cultivars, known collectively as autumn-fruiting raspberries, produce smaller numbers of flowers at the tips of the canes produced during the current season. Blackcurrants on the one hand, and their close relatives red and white currants on the other, also display different flowering characteristics. Red and white currants have a complex ancestry, involving at least three species of *Ribes* and they bear their flowers on the old wood, in clusters at the base of the previous year's shoots, and to some extent actually *on* these shoots. Blackcurrants, derived principally from *Ribes nigrum*, on the other hand, bear most of their flowers on the previous year's shoots; and the management and pruning must, therefore, be directed towards these different ends.

Because of their need for support, in the absence of the general tangle of vegetation that would perform this role naturally, raspberries are usually grown in free-standing rows (although, in limited space, they may be grouped around a central pole or ranged along a fence). Blackcurrants are grown as bushes, but red and white currants, although usually seen as bushes, are very amenable to training as cordons (see p. 187). Any food plant that is grown in rows and needs sun to ripen its produce must be positioned carefully. Rows running north–south will receive more uniform illumination than those running east–west.

Herbaceous borders

I have considered the ecology of the herbaceous border in some detail in Chapter 4 as an example of a layered habitat, but I would emphasise again the most important feature of its gradual change throughout the season. In a comparable natural habitat, the seasonal changes are the result of ecological factors and the changing appearance of the vegetation takes no account of aesthetics. In a garden, however, attractive appearance is the over-riding consideration, and it is the skill of the gardener that governs how successfully he bends ecology to his will to achieve this end. A welcome trend in the herbaceous plant catalogues of the better nurserymen in recent years has been to provide more complete information on spacing distances, flowering season and ultimate height, as well, of course, as colour. An example of the way this type of information can be used to design a border is shown in Diag. p. 168. Nonetheless, I do believe that those who can design a border that is efficient both in function and use and that will not need to be changed and rearranged within a year can probably be counted on the fingers of one hand. For the mere mortals among us, a herbaceous border should be something of a chess-board; you should be able to decide on certain key positions (the Kings, Queens and Castles) with the aid of a good catalogue, a measuring tape and squared paper; but for the Bishops, the Knights, and above all, the Pawns, you will need a notebook and preferably a camera. Look hard at the border at regular intervals during the summer and make notes; see which plants are taller than you expected, which provide unacceptable colour clashes and where there are gaps late in the season because you forgot that, with some plants, the foliage dies down after flowering whereas with others it doesn't. Then each autumn becomes a time for rearrangement, for discarding (never be afraid to remove plants that don't fit) and for buying a few more. After about three seasons, your border should begin to take something approaching a satisfactory shape; but remember that even the plants you intend to leave in the same position will need periodic lifting and dividing if they are not to outgrow their allotted space.

Herb gardens

The area of garden devoted to culinary or ornamental herbs is best thought of as a small short-lived herbaceous border, with rather special problems dictated by the widely varying nature of the plants of which it is composed. I stress short-lived, because most herbs very rapidly outgrow their usefulness, through becoming woody, leggy and far too

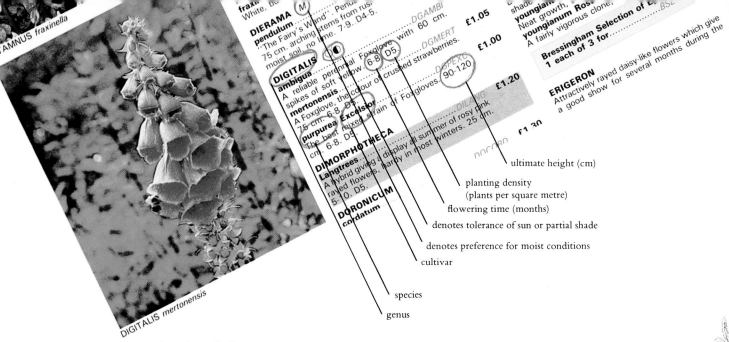

TAMNUS fraxinella

DIGITALIS mertonensis

fraxi...
White, b...

DIERAMA
pendulum (M)
"The Fairy's Wand". Pendu...
75 cm. arching stems from rus...
moist soil, no lime. 7-9. D4-5.

DIGITALIS DGAMBI
ambigua with 60 cm. £1.05
A reliable perennial Foxglove...
spikes of soft yellow. 6-8. D5.
mertonensis DGMERT £1.00
A Foxglove, the colour of crushed strawberries.
75 cm. 6-8. D5.
purpurea Excelsior DSPEXC £1.20
The best mixed strain of Foxgloves.
cm. 6-8. D5. 90-120

DIMORPHOTHECA
Langtrees DILANG
A hybrid giving a display all summer of rosy pink
rayed flowers, hardy in most winters. 25 cm.
5. D. D5. £1.30 DOCOR.

DORONICUM
cordatum

younghum Ros...
Neat growth, Rose...
younghanum clone,
A fairly vigorous clone,

Bressingham Selection of ...BSL...
1 each of 3 for

ERIGERON
Attractively rayed daisy-like flowers which give
a good show for several months during the

ultimate height (cm)

planting density
(plants per square metre)

flowering time (months)

denotes tolerance of sun or partial shade

denotes preference for moist conditions

cultivar

species

genus

The modern herbaceous border is the result of many years development since its early beginnings with Gertrude Jekyll and her contemporaries at the end of the nineteenth century. It is labour intensive and requires great knowledge of plant growth rates, flowering times and requirements and also vision on the part of the gardener if it is to achieve success. In recent years, the more enterprising nurserymen have come to the gardener's aid by providing much more information in their catalogues than was given previously. The illustrations show how, by using the information from such a catalogue, it is possible to draw up a planting scheme that takes account of many plant variables. When such a scheme is put into practice and reaches its ultimate fruition, the result can be a magnificent garden feature that epitomises ecological understanding and ecological principles better than almost any other. (It should be noted that some telescoping of the seasons has been used in the illustration in order to show the flowering of as many of the plants as possible.)

Key to plants used in the border
1. Bergenia cordifolia purpurea
2. Cimicifuga racemosa
3. Lilium henryi
4. Echinops ritro
5. Aconitum napellus 'Bressingham Spire'
6. Delphinium 'Pacific Hybrids'
7. Digitalis hybrids
8. Achillea filipendulina 'Gold Plate'
9. Paeonia hybrids
10. Chrysanthemum maximum 'Snow Cap'
11. Dahlia hybrids
12. Helenium 'Copper Spray'
13. Helleborus orientalis hybrids
14. Euphorbia polychroma
15. Hosta 'Krossa Regal'
16. Geranium 'Claridge Druce'
17. Doronicum 'Harbur Crewe'
18. Gladiolus 'Butterfly cvs.'
19. Phlox 'Pinafore Pink'
20. Ceratostigma willmottianum
21. Polygone affine 'Dimity'
22. Sedum spectabile 'Autumn Joy'
23. Geranium rectum album
24. Lavandula nana alba
25. Aster 'Little Pink Beauty'
26. Hosta sieboldiana elegans
27. Geranium grandiflorum plenum
28. Calamintha nepetoides
29. Alyssum saxatile compactum

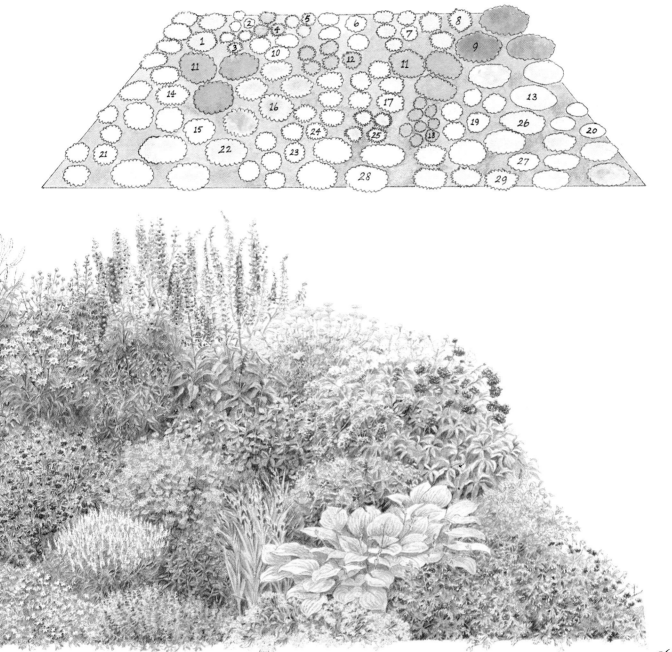

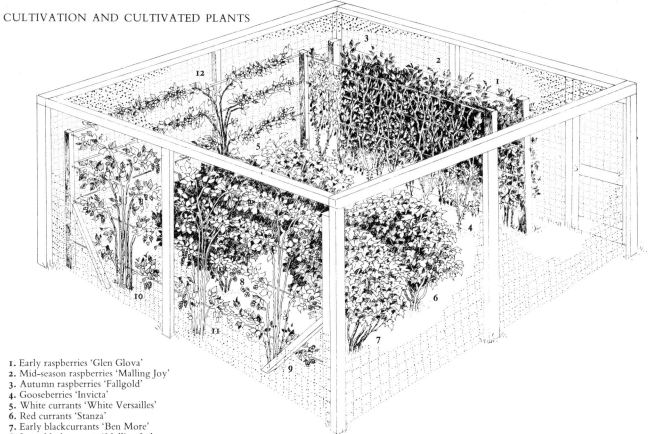

1. Early raspberries 'Glen Glova'
2. Mid-season raspberries 'Malling Joy'
3. Autumn raspberries 'Fallgold'
4. Gooseberries 'Invicta'
5. White currants 'White Versailles'
6. Red currants 'Stanza'
7. Early blackcurrants 'Ben More'
8. Later blackcurrants 'Malling Jet'
9. Blackberry 'Merton Thornless'
10. Loganberry 'LY 654'
11. Tayberry
12. Wineberry (Rubus phoenicolasius)

*Soft fruit are best confined within a cage as this
provides the only really effective protection against
birds, which are likely to cause havoc with both
buds in winter and fruit in summer. The cage
shown includes reliable representatives of all of the
common and a few of the less common fruits
needed by a typical family.*

large. Most are, moreover, very easily rooted from cuttings, and late
summer should be a time for taking these cuttings ready for replanting
in the spring. The herb garden tends to be dominated by rather few
plant families, most notably the Labiatae (mints, sages, thymes, hyssop,
savory, origano, marjoram, basil, and balm, among others); but even
within this family, and indeed within its genera, there are marked variations. Many of the mints, for instance are plants of cool, moist places and,
being generally tall-growing, are best placed towards the back of the
bed (and confined in sunken pots, for their rhizomes are far-reaching
and invasive). The creeping, small-leaved *Mentha requienii* is one exception for nearer the front, however. The thymes are almost all sun-loving
herbs for the front of the border too, but vary considerably in life-form,
some being far-creeping, others fairly tall and bushy, and yet others
extremely slow-growing, rounded and compact. The variegated

thymes display well the generally less vigorous growth and lower frost hardiness that is so often the hall-mark of plants lacking part of their chlorophyll complement.

Shrubberies and woodland gardens

I have discussed woodlands in some detail earlier in the book, and the garden woodland differs from the natural woodland primarily in containing species from a wide range of natural plant associations. Perhaps more than most types of plant, many trees are adaptable to environmental conditions outside their optima and different from those they experience in their natural environment. In gardens, at least a part of their success is due to their having virtually no plant competitors once they have become established. I repeat my advice, nonetheless: never underestimate the vulnerability of young trees to competition from weeds.

The shrub border is generally a fairly easy border to manage, once attention has been paid to the differing growth rates of the various component species and once the principles of pruning have been understood. Nonetheless, in actual choice of species, the shrubbery can sometimes create difficulties. In the wild state, a woody plant of shrub size (and there is no rigid division between a tree and a shrub, although the former is generally thought of as exceeding about 4 m in height on a single stem) may be a dominant member of a habitat of mostly low-growing plants (gorse (*Ulex* spp.) on heathland, for example), or a member of one of the understories of a layered woodland community. Hence, whereas the gorse is a sun-loving plant with a high light demand, rhododendrons, camellias and mahonias, for instance, will only give of their best in a partially shaded situation. Mulching a shrubbery (and I include here the rose garden) is a particularly efficient way of assisting moisture retention in the soil, because the mulch is unlikely to be disturbed by constant cultivation; but if using bark for this purpose, care should be taken to ensure that it is obtained from a reputable source and free from any contamination with honey fungus.

Alpine beds and rock gardens

I have described the mountain-top environment as probably the most difficult to mimic in a garden; there is no way to reproduce at low altitude all the features of high insolation, strong winds, frequent and heavy rainfall, low humidity, thin soil, free drainage and extremes of temperature that are the hall-marks of alpine plant communities in the wild.

Most rhododendrons, unlike many of their close relatives, azaleas, are shade-loving shrubs and are ideal for planting under trees on acid soils where the canopy provides the cool, moist conditions in which they produce their most effective flowering.

The choice of species for growing outdoors at low altitude is probably limited more by the propensity for many of them to rot in the high humidity than by any other factor and these, if they are to be grown at all, really must be accommodated in a purpose-managed, well ventilated alpine greenhouse.

When constructing and planting a rock garden, two points should be borne in mind. First, unlike almost any other garden feature, a rock garden never looks pleasing and satisfying unless it truly resembles its natural counterpart. And this is exceedingly difficult to achieve. One of our most famous British public gardens boasts a rock garden containing many thousands of tons of stone and constructed at vast expense, but which is more akin to the front of a Greek temple than an alpine mountainside. The rock garden must be sited in such a way that a natural rock outcrop *could* arise there, and should preferably be of a rock similar to the local geology. A mere heap of soil with rocks stuck all over its surface is a laughable wart on the face of any garden, what the greatest of all rock gardeners, Reginald Farrer, described as an almond pudding. The second feature to bear in mind is that the rocks in a rock garden do not exist for ornament; they are there to enable you to grow both the plants that can tolerate the heat reflected from the side of the rock that faces the sun, as well as those that must have the shade that its opposite side provides. Choose your plants carefully, and plant them carefully.

Lest I have put you off growing alpine plants by suggesting that you need either a specially built greenhouse, an almost bottomless pocket to buy stone, and the strength of Samson to put the stones together, please read on. Alpines can be grown very successfully and very pleasingly with none of these; they can be grown in troughs and in hollow walls. In raised 'table' beds, where no attempt is made to resemble anything natural, a most attractive well-drained garden feature is produced that can be home to some of the smaller, more compact alpines; many of the mossy saxifrages look exquisite grown in this manner. But do avoid those creeping species (of dwarf *Phlox* and *Campanula* especially) that can subjugate a small alpine trough within months.

Pools and water gardens

Once a gardener has enjoyed the presence of water in his garden he will, most assuredly, never be without it again. Few gardeners have the benefit of a natural pool or water course; but when constructing a garden pool, it is important to decide whether you want a creation that resembles the

Waterlilies are the cream of most people's garden pools; but their selection requires care. They vary greatly, not only in flower colour but also in the rapidity and extent of growth, and in preferred depth of water (see Table 16).

natural or which is unashamedly artificial; those that are attempts at compromise never seem wholly satisfactory. The water itself is treated in more or less the same manner in both approaches, but the natural pool extends into the bog or marginal garden and enables the planting of a range of plants suited to this very specialised niche. What I have called the unashamedly artificial pool has a hard edge, and any planting wholly out of the water is confined to pots and tubs arranged around the periphery.

There are certain essentials that should dictate the siting of a pool. Clearly, it should be on a level site. If it is at the base of a steep slope, water draining into the pool will cause it to overflow – which is excellent if you *do* plan to have a bog garden, but infuriating if you want a formal pool and hope to sit close to the water's edge. Sunlight is also essential, for there are almost no surface-living water plants that are shade tolerant, and water lilies especially, the cream of most people's pools, really need six or more hours of direct sunshine each day. While organic matter breaks down at varying rates in soil, and produces chemical by-products as it does so, these are normally of no significance. Organic matter in a pool is a very different matter, however, for it decomposes quickly, with two consequences. First, it releases sulphur-containing gases that can be harmful to fish and water plants (especially if they are trapped beneath the surface ice in winter); thus, a build-up of fallen leaves is the quickest way to render the water of a pool foul and unproductive. Second, when the decay of organic matter in a pool takes place in the basal soil, it rapidly depletes the soil of nitrogen, in addition to releasing noxious gas. So pools should never be sited close to overhanging trees, and neither manure nor compost should ever be added to the soil in which water plants are grown. For those now conjuring up visions of natural lakes and pools, fringed with overhanging alders and willows, and wondering how they ever thrive, it is important to maintain a sense of proportion. In relation to the volume of water in a natural lake, the addition of the decaying leaves is fairly insignificant (although water at the edge of even a large lake can be relatively deficient in plants and fish, and fairly foul-smelling too).

The ecology of aquatic habitats is, of course, a large subject in its own right and I can no more than touch upon a few other features associated with garden pools. A fountain is, to many gardeners, an integral part of the formal pool (but not, I hope, of the more natural one), and it can play a small part in improving the oxygenation of the water. A large fountain can be counter-productive, however, in bringing about over-cooling of the water with consequently poor growth of plants and fish. In so closed a community as a pool, it is essential to create a correct balance between the various inhabitants (fish, plants and snails) and the volume of water in the pool. Remember too the different life-forms that exist among water plants, and try to achieve the correct balance between them. Table 16 shows examples of some types and quantities of water plants, fish and snails with which to stock pools of different sizes.

Lawns

A lawn is the extreme among those gardening habitats that are, from choice, monocultures. It differs, therefore, even from natural grasslands which, as we have seen, normally contain a mixture of species, with grasses as the dominant members of the community. One of the most taxing and continuing problems of lawn management is in the striving to maintain this monoculture by weed suppression. Although it may be possible to eliminate some of the rosette-forming lawn weeds, such as dandelions and daisies, by mechanical means (digging them out with a 'V'-shaped tool), it is impossible to eradicate creeping weeds like clovers (*Trifolium* spp.) or creeping speedwell (*Veronica filiformis*) in this way. If these are to go from your lawn, a modern selective weedkiller is needed, based for example on such chemicals as dicamba, mecoprop or MCPA; ioxynil, a fairly reliable stand-by for the control of weeds unresponsive to other chemicals, has recently been withdrawn. Once an extensive area of weed growth has been removed from a lawn, the virgin areas of bare soil provide ideal conditions for other weeds to re-colonise unless they are quickly sown with grass seed; weed seeds can germinate, and the resultant seedlings grow, much more quickly than the existing grass can invade the bare areas.

The choice of wild grass species for turfing a lawn was discussed on p. 132, but seeding represents a cheaper, if somewhat slower way of achieving a high quality lawn. There now exist many hybrid grasses, bred for specific horticultural uses; and the commercial groundsman has a wide range of mixtures from which to choose, including different blends for tennis courts and golfing tees, putting greens and croquet lawns, rugby and football pitches, fairways and cricket outfields. Gardeners have a smaller range, but most seed companies offer at least the two basic mixtures of 'lawn seed with rye grass' and 'lawn seed

Table 16 Some suggestions for quantities and types of plant to combine with fish when stocking pools of different sizes

Quantities		Small Pools (c. 1.75 m × 1.25 m × 30 cm deep)	Medium Pools (c. 3 m × 1.75 m × 45 cm deep)	Large Pools (c. 4.25 m × 3 m × 55 cm deep)
	Marginal Plants	6	8	20
	Water Lilies	1	2	4
	Other Pond Plants	1	2	6
	Floating Plants	4	7	20
	Oxygenating Plants	10 bunches	25 bunches	80 bunches
	Fish	10 (8 cm long)	20 (8–10 cm long)	25 (12–15 cm long)
	Scavenging fish	1	3	5

Types

MARGINAL PLANTS

Water plantain *Alisma plantago-aquatica*
Umbrella rush *Butomus umbellatus*
Marsh marigold *Caltha palustris*
Bog iris *Iris laevigata*
Bogbean *Menyanthes trifoliata*
Water forget-me-not *Myosotis palustris*
Pickerel weed *Pontederia cordata*
Arrowhead *Sagittaria sagittifolia*
Brooklime *Veronica beccabunga*

WHITE WATER-LILIES

Alba	v
Marliacea albida	m
Odorata alba	s
Pygmaea alba	t

PINK WATER-LILIES

Colossea	v
Marliacea carnea	m
Pink Opal	s
Laydekeri rosea	t

RED WATER-LILIES

Conqueror	v
Escarboucle	m
James Brydon	s
Ellisiana	t

YELLOW WATER-LILIES

Col. A. J. Welch	v
Marliacea chromatella	m
Sunrise	s
Pygmaea helvola	t

OTHER POND PLANTS

Water hawthorn *Aponogeton distachyum*
Arrowhead *Sagittaria natans*
Golden club *Orontium aquaticum*
Fringe lily *Villarsia bennettii*

FLOATING PLANTS

Fairy fern *Azolla caroliniana*
Frogbit *Hydrocharis morsus-ranae*
Water fern *Stratiotes aloides*

OXYGENATING PLANTS

Canadian pondweed *Elodea canadensis*
Hornwort *Ceratophyllum demersum*
Milfoil *Myriophyllum spicatum*
Water violet *Hottonia palustris*

Key to sizes of water lilies:

v = very vigorous	Surface spread 1.5 m diam	Water depth 30 cm–1.5 m
m = moderately vigorous	Surface spread 1 m diam	Water depth 20–60 cm
s = small	Surface spread 60 cm diam	Water depth 15–45 cm
t = tiny	Surface spread 30 cm diam	Water depth 10–30 cm

without rye grass'. The magic ingredient, rye grass, refers to a species or, more probably, one of the innumerable hybrid grasses belonging to the genus *Lolium*. Rye grasses are quick and relatively tall-growing (although some dwarfer hybrids are now available), fairly coarse-leaved, and can withstand wear and tear. Because of their large leaves and relatively lax habit, they do not respond well to close mowing and never produce a very closely-knit turf. A mixture containing rye grass should be the choice for any lawn that is to withstand being trampled, walked over or played on. One seed company refers to those mixtures containing rye grass as 'back lawn mixtures' and those lacking it as 'front lawn mixtures', and this fairly well sums up their respective uses. But what of the ingredients in addition to the rye grass? The bulk of the remainder will usually be species or hybrids of fescues (*Festuca* spp.) with greater or lesser amounts of *Poa* and *Agrostis* species.

More recently, there have been introduced to the garden market grass as 'back lawn mixtures' and those lacking it as 'front lawn mixtures', such evocative names as 'Summer Glade' and 'Greenshadows'. These usually contain a higher proportion of those forms of *Festuca rubra* that are tolerant of a light shade. Unfortunately, the grasses tolerant of higher shade, such as *Poa nemoralis*, have too much an upright, rather than a creeping habit and cannot therefore tolerate mowing.

Supports for climbing plants

The climbing habit, because of having insufficient physical stem support to remain free-standing, has been adopted by a wide variety of plants representative of many different families. In gardens there are certain groups that outweigh all others in importance; roses, *Parthenocissus*, *Vitis* and related vines, *Clematis*, ivies and honeysuckles most notably. We have British representatives of almost all of these and a gardener needn't look far, therefore, to see how these plants grow naturally. They obtain their support from other plants, from rocks, sometimes from the ground if they can creep and scramble as well as climb, and, very commonly, from themselves; a mass of bramble, for instance, may be three metres high, but quite hollow underneath. The type of support to provide for these plants in gardens depends to some extent on the manner in which particular species have elected to climb. This ranges from the scrambling habit, largely dependent on backward-facing prickles (represented by roses), through grasping tendrils which may be either modified leaf stalks (as in *Clematis*) or modified stems (as in vines),

Climbing plants use several different ways of supporting themselves on other vegetation or, in gardens, on artificial supports. Roses (right) for example, have backward-facing thorns that act like grappling irons, while the runner bean (far right) obtains a much tighter hold with twining stems.

to extremely efficient modified aerial roots (possessed by ivies and a few other so-called self-clinging species). The latter will grow over almost any surface of their own volition and can, indeed, be positively destructive on old mortar and old bricks. All others need some help in the shape of wires or trellis is they are to be persuaded to cover a wall or other 'unnatural' surface. Never forget the potential of climbing plants to adopt their natural habit however, and use them to obscure or embellish old tree stumps or, indeed, entire trees.

Climbing plants vary greatly in their rapidity of growth and it is essential to choose species and cultivars appropriate to a particular situation. Table 17 lists some of the more readily obtainable climbing plants, representative of a range of habits, suitable for the more common garden habitats. Certain features of the environment in which these plants grow present particular problems. Perhaps the biggest is that the base of a wall or fence, where they are rooted, is slightly sheltered from the rain and inherently very dry: additional watering and feeding will be necessary, especially in the early years. The dryness of the atmosphere, moreover, will often render climbing plants especially prone to powdery mildew, although the shelter from wind and cold that a wall affords does mean that a plant insufficiently hardy to be grown exposed in the garden, may survive if trained up the (correct) side of a house. I shall say a little more about the principles behind training on p. 183, but it should not be forgotten that the shelter of a wall affords the opportunity to grow not only climbing plants but also those usually known as wall shrubs. Some of these, like *Pyracantha*, can grow elsewhere in a 'normal' form and are also amenable to being trained into an attractive and more or less two dimensional form up a wall; but others, like *Pittosporum*, *Fremontodendron*, *Ceanothus* and, of course, many Camellias, may be too tender to survive as free-standing individuals in the garden without the protection of a wall.

Protected growing

In Chapter 2, I described the way that a greenhouse enhances some features of the climate and affords protection from the effects of others. Accepting, therefore, that the benefits conferred by a greenhouse are those of facilitating the growing of plants that are insufficiently hardy for the rigours of the great British outdoors (or can only tolerate the outdoor life at certain times of the year), or of inducing plants to grow, flower and fruit prematurely, I shall now examine the practical constraints of the environment within the greenhouse itself.

Diag. p. 43 shows a greenhouse containing a wide range of different types of plant which originate in very different ecological communities and have very differing microclimatic requirements. Tomatoes, originating in South America, always thrive best when wet at the roots but fairly dry above ground; an irrigated garden in the hot, dry Mediterranean summer always seems to suit them ideally. Cucumbers, on the other hand, originated in the humid tropics or sub-tropics (having been cultivated since antiquity – no-one seems very sure quite where) and luxuriate in the steamy heat that merely causes tomatoes to rot. And what of the other familiar greenhouse plants? Most of the remaining vegetable fruits (peppers, okra and aubergines, for instance) fall somewhere between the two extremes; ornamental flowering and foliage pot-plants display every combination of requirements; grapevines seem to have similar demands to tomatoes; and alpines positively insist on the peculiar conditions prevailing at altitudes from around 1,000 m up to the snow line.

Logic suggests that to house all these plants in the same greenhouse is to court disaster; but this is only true up to a point, for this is one of those areas of gardening where theory doesn't always match up with practice, and where the standards followed in commercial horticulture are not necessarily applicable to gardening. Commercial growers and nurserymen, requiring the maximum possible return from their plants, will not mix the various types together. In a garden greenhouse, nonetheless, even with plants of such widely contrasting origins, it is possible to achieve a compromise, by adopting two complementary approaches. The first is to choose plants within each group that are not demanding of extreme requirements; and the second is to subdivide the greenhouse. Among alpines and pot plants, you should certainly avoid any requiring, constantly, either extreme cold or extreme heat.

Few tomato varieties suitable for a British greenhouse will tolerate very high humidity. Varieties suitable for tropical conditions do exist, but they will not grow well in our climate, and I do not believe they are worth growing simply to facilitate coexistence with cucumbers. Much easier and better is to choose those cucumber cultivars more tolerant of lower humidity; in my experience 'Conqueror' and its relatives are the best of these, and will grow well in an unheated greenhouse. But you will find it almost impossible to grow alpines in an undivided greenhouse (even if it is unheated) containing tomatoes and cucumbers (because of the enhancement of the humidity by the water given off through their leaves). And it is when you begin to introduce early-season heating that trouble really begins: your alpines will rot away almost overnight, and an insulated partition becomes essential.

In theory, I suppose, no space is too small to subdivide, but a partition across a greenhouse smaller than about 2×2.5 m will create one almost unworkably small compartment if the other is to accommodate more than two tomato plants. Given that the partition itself must be insulated, and yet capable of allowing you to gain access to the inner compartment, trying to achieve the compromise I have discussed is not really worthwhile unless the greenhouse is at least of this size.

Table 17	Some widely available climbing plants for different environments	
Plant	**Valuable features**	**Especially suitable for . . .**
Actinidia kolomikta	Creamy-white, pink flushed leaves	South or south-west wall for best leaf colour
Campsis radicans	Orange-scarlet trumpet flowers	Full sun; best over old buildings or tree stumps
Cissus antarctica	Small, glossy green leaves	Conservatory or house; sun or shade
Clematis alpina cvs.	Early; bell-like flowers; moderate vigour	Low walls, tree stumps; north facing sites
Clematis macropetala cvs.	Flowers after *alpina*, similar flowers and vigour	Low walls, tree stumps; good in containers
Clematis montana cvs.	Spring flowering; very vigorous; a few are scented	Old buildings, old trees; best in sun where they can have free rein
Clematis orientalis	Late summer; yellow, lantern-like flowers	Best growing through old trees; too vigorous for restricted sites
Clematis tangutica	Late summer; yellow flowers, smaller than *orientalis*	As *orientalis*; combination of flowers and seed heads most attractive
Clematis viticella cvs.	Summer; moderate vigour; some with very deep red flowers	Best trained into a small tree or over low-growing shrubs
Clematis hybrids	Generally large flowered; range through the summer months	Suitable for wide variety of sites, but many have best colour facing north
Hedera canariensis cvs.*	Very large leaved ivy; best forms are variegated	Walls, fences or as individual specimen plants in courtyards
Hedera helix cvs.*	Common ivy; wide range of leaf colour and shape	Self-clings to almost anything (but avoid old bricks); tolerates deep shade
Humulus lupulus 'Aureus'	Golden Hop; yellow leaves and drooping greenish flowers	Vigorous in full sun; good over a pergola
Hydrangea petiolaris	Climbing hydrangea; self-clinging; white flowers	Valuable for a north wall but dull in winter
Jasminum officinale	White-flowered, scented summer jasmine	Well supported, it will grow almost anywhere, but best near to house for scent
Lonicera japonica cvs.*	Very vigorous; fragrant summer flowers	As cover for old buildings, trees and other sites where they can have free rein
Lonicera periclymenum cvs.	Common honeysuckle; cvs. vary in flower colour and scent	As *L. japonica*, but most gardens should find room for one
Parthenocissus spp.	Four common spp., incl. Virginia creeper; autumn colour	Good on a north wall; glorious in autumn; dull in winter and flowers insignificant
Passiflora caerulea	Passion Flower; exotic and dramatic flowers; vigorous	Need a warm sheltered wall in most parts of the country
Polygonum baldschuanicum	Russian Vine; white flowers; incredibly vigorous	Will grow almost anywhere but extreme vigour means it should be used as rapid cover only
Rosa spp. and cvs.	Immense range of colour, vigour, flower type and perfume	Species roses are best for less formal situations such as trees and old walls
Solanum crispum	Vigorous; potato-like purple-blue flowers; good on lime	Less hardy and less vigorous than *P. baldschuanicum* but can be used similarly
Vitis coignetiae	Best ornamental vine; vigorous; big leaves; autumn colour	Large walls and fences, tall trees; needs some support for the tendrils
Wisteria spp.	Majestic, long-lived; pendulous purple or white flowers	Best flowers in sun; can be magnificent trained over a pergola or into a very big old tree

* *Consistently evergreen (although others may retain some leaves in warm sites and mild winters)*

Soil-less growing systems

A few plants grow naturally in environments where there is little or no soil; their adaptation to abnormal conditions must be considered in cultivation. Alpines, even those that naturally inhabit screes, usually have their roots in at least the modicum of soil that exists in these environments; as described on p. 171, these conditions are usually recreated by using a very well-drained growing medium containing a high proportion of grit. Plants that display the epiphytic life form and grow on the surface of other plants (trees especially) may also have their roots in a small cleft or crevice of thick bark in which soil particles have accumulated. Some have roots embedded in the bark itself, however, and when such plants are cultivated (most, being tropical, are usually considered greenhouse subjects), they usually thrive best in a soil-less compost in which the bulk of the medium is either peat, pulverised bark or similar organic material.

In fact, many plants that grow naturally in soil will also thrive in such soil-less substances provided certain requirements are met, and indeed there are several advantages in cultivating them in this way. Peat, bark and similar materials can for present purposes be considered sterile, and thus present none of the pest and disease problems attendant on the use of soil in the abnormally warm conditions of a house or greenhouse. They are relatively clean (an obvious advantage indoors), are not prone to capping and compaction when used in a pot or other type of container, are immeasurably more uniform in composition than soil and much less dense (both obvious advantages for the commercial plant raiser seeking standardisation of his environment and ease of transportation). Peat and bark have one major additional attribute: they contain almost no nutrients. The technology of blending chemical fertilisers with them in order to produce a balanced growing medium still has some way to progress, but it is nonetheless now possible to achieve consistently uniform results for a limited time, after which the nutrients become exhausted, and liquid fertilisers must be employed. The classic vehicle for soil-less growing systems in gardens is the plastic growing bag, containing a peat-based compost for raising tomato or other annual plants, either outdoors or in a greenhouse. It is essential that you pay careful attention to the manufacturers' advice regarding supplementary feeding, however, and to appreciate that watering is not always easy to judge, for the compost can dry out or, conversely, become waterlogged, very rapidly. At the end of the season, the nutrients in the bag will be exhausted and it is pointless attempting to raise further plants in it. Much better is to dig the compost into the garden and make use of its bulk as a soil amendment.

Miniature gardens

The subject of soil-less growing media leads neatly onto the ecology of miniature gardens in general, a term that includes not only growing bags, but also tubs, window boxes, hanging baskets and other containers. Outdoors, these containers are usually filled with soil-based rather than soil-less composts, and thus the fertiliser requirements of plants in them are somewhat less critical; but it should be appreciated that some supplementary feeding will be necessary when many plants are not only crowded together, but also expected to produce a superfluity of flowers or fruit. Weeds are not normally a problem in a container; it is merely the competition between the plants themselves that must be considered. Pests and diseases will spread very rapidly in these closeted conditions and must be watched for assiduously. Moreover, my remarks about under- and over-watering are also applicable; the provision of adequate drainage is essential if the plants are not to suffer in the same way as those growing naturally in a soil with impeded water flow. Hanging baskets, one of the most bizarre habitats in which any plants can be expected to flourish, suffer from drying out and nutrient depletion almost more rapidly than any other gardening environment.

One particularly interesting contained growing system is that known as ring culture (left). Used in a greenhouse, this is an alternative to the growing bag in providing a carefully controlled environment in which tomatoes or other plants will crop efficiently; and it has none of the disadvantages in terms of pest and disease problems that arise when the soil within a greenhouse is used. The ring-culture method depends on encouraging the plant to develop two separate root systems, one of which is used for water uptake only and the other for the uptake of nutrients provided in the form of liquid fertiliser. In many ways, I feel this represents almost the extreme in ecological abuse of the plant (matched only by commercial growers who raise lettuces in what amounts to little more than fertilised concrete); but who is to say that such a plant, with its rich, luxuriant foliage and swollen trusses of glorious red fruit is any less dignified than its wild ancestor, scrambling over boulders in the Andean foothills and struggling to produce a few pea-sized tomatoes?

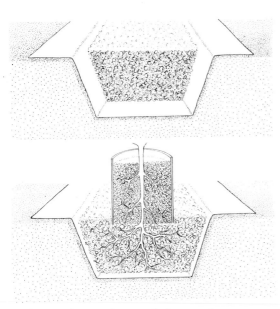

In the ring culture 'miniature garden' system, the plant (usually a tomato) forms one root system in a bottomless pot of compost for nutrient uptake and the other, in a gravel bed lined with plastic sheet, for water absorption.

Pruning and training

Cultivating a plant entails more than merely placing it in a garden and satisfying the natural needs for food and water that removal from its natural environment may have denied it. It also entails employing other techniques to ensure that the plant meets your expectations in terms of appearance or fecundity and, indeed, that as far as possible it will not outgrow its allotted space. There is a need for some bending of nature to your will, and nothing illustrates this better than the greenhouse tomato and the hedge. Left to its own devices, the tomato is a much branched plant, producing buds, and consequently shoots, in the axils of the leaves. The bush tomato cultivars grown outdoors in gardens are of this so-called 'determinate' type. Usually, they produce a fairly large number of fairly small fruit. Over the years, however, it has been decided that big is better for tomatoes, and cultivars have been bred for greenhouse culture that are potentially capable of producing larger fruit. I say potentially, for, in our climate at least, you can't have your tomatoes both big *and* determinate; to produce larger fruit, you must accept fewer fruit, for the root system of the plant is capable only of taking up a certain quantity of nutrient and water. To persuade the plant to produce fewer but larger fruit therefore, the shoots bearing some of the flowers must be removed. This is achieved by the simple expedient of side-shooting. Similar principles are employed in the removal of some of the buds of dahlias, chrysanthemums and other flowers, in order to produce the solitary giants so beloved of some gardeners. In theory, many other types of plant will develop in a similar manner if they are artificially prevented from achieving their natural purpose. Fortunately, however, the single giant pea-pod has not yet caught the horticultural imagination!

The tomato and other plants that I have described have been subjected to a form of pruning, in that parts have been deliberately amputated. The plants that constitute a hedge, left to their own devices, would rapidly become trees or large shrubs and it is pruning in the form of clipping that prevents them from doing so. But in this instance it is more, not less, leafy growth that is required; by removing the apical bud from each shoot the clipping process limits its extension and in most cases encourages other buds to break from lower down and thus produce a more compact and bushy form.

There are yet other reasons for pruning. The same technique may be used to remove dead or diseased wood, to remove branches and shoots that have become so tangled together as to restrict the penetration of

By removing the side shoots, a tomato plant that has a naturally bushy habit can be induced to develop into an upright, single-stemmed form – a more manageable plant in need of staking and with fewer, but better fruits.

light and air to the plant, or to remove dead flowering shoots, either to prevent them from becoming diseased or simply to improve the appearance of the plant.

Pruning is one of the most mysterious of the gardening arts and, on countless occasions, I have been faced with the question 'My so-and-so bush is growing very well and putting on plenty of new growth. How should I prune it?' The answer in eight times out of ten is 'Not at all'. Nonetheless, as there are occasions when a plant *does* make a better garden subject after at least *some* pruning, I have listed in Table 18 a few of the commoner ornamental shrubs with notes on how they should be treated. The more perceptive and better informed among you will see that there is a pattern to this, and it is important that this is appreciated if pruning is to be elevated from the mysterious to the methodical.

The pattern relates pruning time to flowering time; some shrubs form their flowers on wood produced in the previous year and tend to be in bloom early in the season, before the current year's shoots have developed. These may be pruned as soon as the flowers fade. Shrubs that flower on wood of the current year may be pruned between late autumn and the end of February, the latter being preferable. If you follow this general rule, you are unlikely to do very much harm and are unlikely to prune away all the next crop of flower buds. Relatively few shrubs of the first type require a great deal of regular pruning, apart perhaps from the removal of old flower trusses, and a little shaping. Some of the latter group do require more severe and regular treatment to give of their best, but the basic rule should be not to prune hard unless you have checked that this is appropriate. *Amorpha, Buddleia davidii (not* other buddleias), *Hypericum calycinum, Indigofera, Lespedeza,* and *Rhus typhina* (when grown solely for its foliage) are among the few shrubs needing a really hard pruning (to within a few buds of the base) in February.

The same general principles apply to climbing plants, and clematis illustrate this well, for the pruning of this group often causes confusion. The genus *Clematis* can be divided into three groups on the basis of their flowering biology. The first comprises the early flowering species *C. alpina, C. macropetala* and their respective cultivars, and *C. montana* and its cultivars. These should be pruned immediately after flowering, simply to remove dead shoots and to keep the plant within its allotted space. The second group comprises the early, large flowered cultivars such as Dr Ruppel, Nelly Moser, Elsa Spaeth, Etoile de Paris, Mrs

Table 18	General pruning methods to be adopted with some of the commoner ornamental garden shrubs
Shrub	**Pruning system**
Berberis	Thin out old wood in spring after flowering; clip hedges after flowering
Buddleia	Cut back *B. davidii* to four buds from the base in early spring. Shorten shoots of other buddleias after flowering to improve shape
Buxus	Bushes normally need no treatment. Clip hedges in late summer
Calluna	Clip back old flowering shoots in early spring with shears
Camellia	No regular treatment needed, but old, leggy plants can be cut back hard after flowers fade
Ceanothus (spring fls)	As a bush, thin only in spring. Against a wall, cut back to within two buds of previous season's growth after flowering in spring
Ceanothus (summer fls)	Cut back hard to within 2–6 buds (less for weaker shoots) in early spring
Chaenomeles	Cut out old wood, thin out and shorten side shoots after flowering
Choisya	No regular treatment needed. Cut out old wood and shape old plants in spring
Cotoneaster	No regular treatment needed but rejuvenate old bushes by hard pruning in spring; early for deciduous, late for evergreen
Cytisus	Prune each year to shape after flowering, but do not cut beyond previous year's wood
Elaeagnus	No regular treatment needed but thin out and shape in spring; early for deciduous and late for evergreen
Erica	Shape by removing dead flower heads with shears as they fade
Escallonia	No regular treatment needed for bushes, but, against a wall, thin out and shape after flowers fade
Euonymus	Shape only during summer
Forsythia	Against walls, cut out old flowering shoots after blossom has faded. In the open, thin only, every three years
Fuchsia	Large, hardy bushes are best given no regular treatment. Smaller and less hardy plants should have dead growth removed in spring
Hebe (Veronica)	No regular treatment needed but old bushes can be rejuvenated by pruning hard in spring every few years
Hydrangea (mop-head)	Cut out dead flower heads after flowering in very mild areas but leave until spring in most parts of the country
Hypericum	Thin out in early spring. Vigorous types like *H. calycinum* should be clipped with shears
Jasminum	Cut back flowering shoots of winter jasmine to within two buds of the base after flowering. Thin out summer jasmine after flowering
Kerria	Cut out most of the old wood after flowering
Laurus	Prune in spring; hard for old leggy bushes, less for vigorous young ones
Lavandula	Cut back old flowering shoots in late summer
Magnolia	No regular treatment needed. Shape as necessary after flowering or in early autumn
Mahonia	No regular treatment needed but remove long, non-flowering shoots and old wood in spring
Olearia	Remove dead flowers in late summer
Philadelphus (hybrids)	Remove old flowering shoots and thin out generally after flowering. Rejuvenate old plants by pruning hard in spring
Potentilla	No regular treatment needed, but cut out old wood in early autumn after flowering
Pyracantha	No regular treatment needed for bushes in the open. Against walls, shorten secondary shoots in summer and spring
Rhododendron	Very carefully remove dead heads after flowering. Rejuvenate old bushes by hard pruning in spring
Ribes	No regular treatment needed, but old bushes can be rejuvenated by hard pruning in spring
Spiraea	Cut out old flowering shoots after flowering (spring flowering types) and prune hard in early spring (late summer flowering types)
Syringa	Remove dead flower heads after flowering. Remove suckers annually. Rejuvenate old bushes by pruning back growth by up to one half in spring
Tamarix	Prune small bushes of late summer flowering forms hard in early spring. Large bushes should merely be shaped
Viburnum	No regular treatment needed although *V. tinus* responds to shaping in spring
Weigela	No regular treatment needed, but remove leggy shoots after flowering
Wisteria	Shorten young shoots back to six buds in July and then to two buds in December

Cholmondeley, Niobe and The President, together with the Jackmanii group, and the mid-season large flowered cultivars such as Beauty of Richmond, Duchess of Sutherland, Marie Boisselot and W. E. Gladstone. These should have dead or feeble shoots cut out in February or March and other shoots cut back to a pair of strong buds. The final clematis group includes the *C. viticella* cultivars and all the late flowering species such as *C. campaniflora*, *C. flammula*, *C. orientalis*, *C. rehderiana*, *C. tangutica*, and *C. texensis*. These should also be pruned in February or March, but much harder, with all the previous season's growth cut back to a strong pair of buds just above the base. Although this sounds complicated, it will be seen, in fact, to follow the guidelines relating pruning to flowering time that I mentioned above, with the same additional comment that some plants need a hard February pruning and some don't.

The technique of pruning is the means to an end, the end being training, the name generally given to the persuading of a plant to grow into a chosen shape and form. It is inappropriate here to describe the many methods of training available for fruit trees, climbing plants, shrubs and trees ranging from the miniaturised trees of bonsai form, through the outlandish results of the topiarist's art, to the often bizarre shapes into which apple and pear trees have at times been forced. Diag. p. 186 illustrates some of the variations that are possible, but all depend ultimately on removing some buds on a plant in order to induce other, dormant buds to break into life.

Some fruit tree training systems
Many, wonderful and sometimes weird are the shapes that fruit trees can be induced to adopt by the expedient of removing some of their buds and shoots to encourage the extension of others. The underlying objective is to improve cropping by the development of fewer but better fruits and also to aid picking.

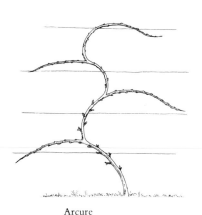

Arcure

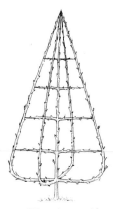

Winged pyramid

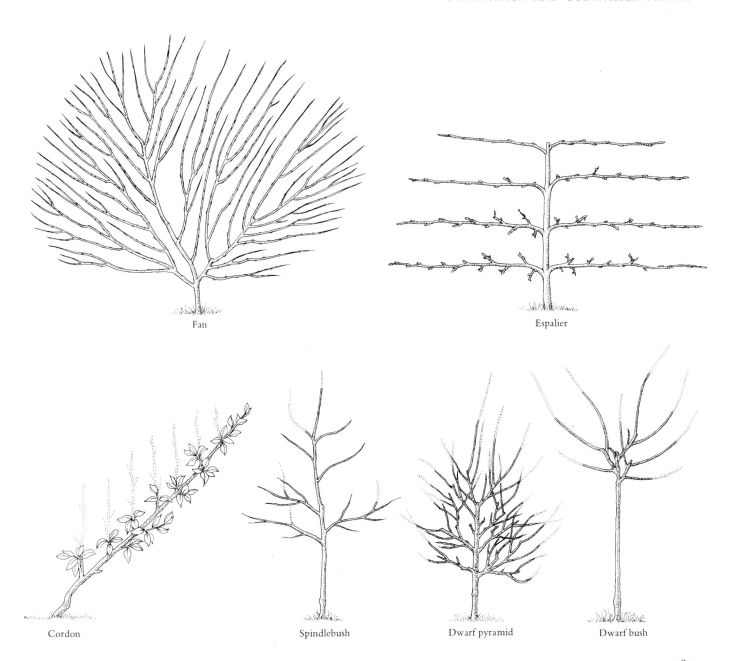

Fan

Espalier

Cordon

Spindlebush

Dwarf pyramid

Dwarf bush

Wind pruning.

We seem, with topiary and bonsai, to have strayed a long way from the natural growth of plants in an ecologically balanced habitat, but a little thought will reveal that this is not so. We have all seen the mis-shapen trees on the windy sea shore, forced to grow in one direction because the drying wind has killed the buds on the sea-ward side. And it is said that the early development of bonsai in Japan was through the observation of miniature trees growing in rocky clefts where lack of nutrients and soil brought about a natural form of root pruning. We may be able to do some decidedly strange things to plants in cultivation but, ultimately, we can only use the resources of the plant itself and adapt for our own, often curious ends a capability that has evolved for some quite different ecological purpose. With this observation in mind therefore, I shall now describe the ways in which some of our common-est garden plants differ from their wild ancestors and how cultivation and breeding have made use of particular aspects of their natural varia-tion and ecological behaviour.

Apple

The wild crab apple (*Malus silvestris*) is one of three species (*M. pumila*, a smaller tree from the Caucasus and Turkestan, and *M. baccata* from Russia being the others) that have contributed to the long ancestry of the modern cultivated apple. Although crab apples are usually thought of in Britain as hedgerow trees, all *Malus* species are very much at home in the understory of broad-leaved woodland, and it is the experience of many gardeners that apples are more shade tolerant than they are usually given credit for. *M. pumila* is generally thought to have occurred somewhere in the ancestry of many of the dwarfing apple rootstocks, while *M. baccata* has been valuable for its winter hardiness and resistance to scab. The modern apple, divided loosely into dessert, cooking and cider forms with increasing acid content, is genetically so complex that pips sown from any particular fruit are most unlikely to give rise to trees that have any value for fruit production. But it should not be forgotten that when a retired brewer called Richard Cox did just that in 1825 and sowed two pips from an apple called Ribston, one produced Cox's Pomona and the other, Cox's Orange Pippin!

Pear

The pear has much in common with its close relative the apple, the wild species *Pyrus communis* from which most modern pears are derived being a woodland and forest tree found over much of Europe and northern Asia, including Britain where it was probably introduced and is now found mainly in hedgerows. Pears tend to mature more slowly than apples and are now almost invariably grafted onto rootstocks of quince (*Cydonia oblonga*) on which they produce smaller trees and earlier fruit of better quality. Most clones are self-sterile and although Conference in particular will produce good fruit when grown as a single tree, it seems usually to achieve this through parthenocarpy (the formation of seedless fruit without fertilisation) rather than self-fertility.

Plum

Like apples and pears, the ancestors of the plum were small woodland trees. The cultivated plant is usually called *Prunus domestica* and was introduced into Britain before Roman times and still occurs as a hedgerow tree. Its two forbears also occur wild here, the cherry plum or myrobalan *P. cerasifera* which too was introduced, and the native *P. spinosa*, the sloe or blackthorn. The original natural cross between the two seems to have occurred somewhere in that hot-bed of plant origins, the Caucasus. The fruits of *P. cerasifera* are yellow with red pigmentation, those of *P. spinosa* yellow with blue pigmentation; hence the enormous range in colour now found among the various groups of plums and gages. Damsons are a separate group, probably derived from a plant generally called *P. insititia*, the bullace. As with the other tree fruits, plum clones are usually grafted onto a separate rootstock, now most

commonly a stock derived from the bullace and called St Julien A; there are no really dwarfing rootstocks for plums.

Strawberry

The strawberry has a complex history. The common British wild strawberry, *Fragaria vesca*, a native plant that occurs widely in open woodland on chalky soils throughout Europe, Asia and eastern North America, has in fact contributed only to the development of the small fruited 'perpetual' alpine strawberry, which betrays its origins by being very much more shade tolerant than the larger fruited garden strawberry, and makes a useful plant for the front of a partly shaded border. The modern large fruited plant is derived from a cross between the North American Virginian strawberry *F. virginiana* and the Chilean strawberry *F. chiloensis* which occurs widely in mountainous areas in both North and South America. Longer fruiting periods may be expected with new cultivars in the future by using Californian types whose fruiting season is limited only by temperature and not, as with our present British strawberries, by day-length.

Raspberry

I have already referred to the ancestry of the raspberry; it is a good example of a plant that, in the wild, is usually a species of woodlands or, at most, of woodland clearings, but whose derivatives benefit greatly from exposure to sunshine for fruit ripening. It is also a fruit that, until quite recently, was derived from a single species, *Rubus idaeus*, a common British native plant. The North American *R. strigosus*, yet another plant of forest clearings, has been the most important additional contributor to the development of modern cultivars.

Gooseberry

Like the raspberry, the gooseberries grown in gardens were until recently derived by selections from a single species, *Ribes grossularia*, arguably a native British plant. Interestingly, these wild gooseberries are often found growing in fairly damp places in woodland – an indication that they will benefit from additional watering in gardens, particularly at the time the fruit are swelling. The use of related species of *Ribes* in gooseberry breeding in recent years has been mainly to impart a measure of disease resistance, and to explore the feasibility of developing thornless types.

Currants

There are three types of garden currant, the black, derived from *Ribes nigrum*, and the red and white, derived from three or more wild species, *R. rubrum*, *R. petraeum* and *R. vulgare*, the three having differing importance in different cultivars. Both *R. nigrum* and some forms of *R. vulgare* occur wild in Britain, especially on alkaline soils, and it is of interest that both currants and gooseberries (and, incidentally, blackberries) are more satisfactorily grown on chalky garden soils than are raspberries which prefer more acidic sites and show marked chlorosis when on lime.

Brassicas

Brassica is, literally, the generic name of what is numerically the biggest group of vegetables in British gardens and, if farm crops like kale and oil-seed rape are added, the largest component of British agriculture after cereals. They are conveniently divided into two main sub-divisions: the cole crops, comprising cabbages and savoys, Brussels sprouts, cauliflowers, broccolis and calabrese, and curly kale; and the root crops, swede and turnip. A few oddments, like Chinese cabbage (actually a close relative of the turnip) and kohl-rabi (botanically one of the coles) make up the whole. Radishes are not brassicas, but belong to the related genus *Raphanus* with which *Brassica* can hybridise. Varying so much in overall form, cultivated brassicas have been selected for particular development of buds (cabbage, savoy and Brussels sprout), of flowers (cauliflower and broccoli), of leaves (kale), of stems (kohl-rabi), or of roots (turnip and swede).

With so diverse an assemblage, it is not surprising that the genus has a long and complex history. A cabbage-like plant is probably the ancestral form of the coles, and the Brussels sprout the most recently developed member. *Brassica oleracea*, the wild cabbage, is a common plant of sea cliffs over much of Europe, including Britain. That the cliffs are often of well-drained chalk and limestone soils explains the requirement of brassicas for at least *some* lime in the soil, and may be the reason why wild brassicas seem to suffer little from the clubroot disease that creates such problems on wet and acid soils in gardens.

Lettuce

The cultivated lettuce is derived from *Lactuca sativa*, a loose-leaved, rosette-forming and sun-loving plant probably originating, like so many garden plants, somewhere in western Asia. Imagine a plant rather like a large dandelion and envisage generations of selection for forms having increasingly less of the rosette habit and more and more of an upright shape and you will have an impression of the relatively straightforward development of the modern lettuce. Cultivated forms range from the very upright cos to the softer, more rounded butterhead, and include types with varying degrees of red pigmentation (so-called red lettuces).

Carrot

Any gardener whose carrots have run to flower and seed will immediately see that the plant's botanical affinities lie with cow-parsley and hogweed in the family Umbelliferae. *Daucus carota*, the wild carrot, has a widespread distribution throughout much of

Europe and Asia, including Britain, where it is found most frequently in dry, grassy places on alkaline soils near the sea. It often accompanies wild cabbages on the cliff tops. The wild plant is only recognisably a carrot if uprooted (an illegal act these days) and, even then, the roots are a pale, slender replica of the garden forms. Selection during cultivation has been for plump, orange roots (although yellow and even white-rooted carrot cultivars do occur), but the requirement remains for deep, light soils that the roots can penetrate easily.

Onion

The genus *Allium* in the family Liliaceae includes onions and their allies, but also embraces over 500 other species, many of which are, incidentally, somewhat neglected ornamentals. Virtually all have the characteristic onion smell and flavour and it is for this and for such features as tolerance of cold, rather than general appearance, that breeding and selection have been made. Many species of *Allium*, including *A. cepa*, the onion, originated in Western Asia where they occur as plants of dry, rocky places; and even the cultivated forms are tolerant of a wide range of soil types. The onion is a good example of a plant for which tolerance does not equate with luxuriance and, in gardens, they reach their full potential in a rich, well-manured soil. The other edible members of the group differ slightly in flavour and in growth habit and include the tufted chives (*A. schoenoprasum*), the perennial bulb onion known as the shallot, which apparently arose from *A. cepa* at some time during cultivation and does not appear in the wild at all, garlic (*A. sativum*) with a cluster of bulbs, the leek (*A. porrum*) which has a scarcely swollen bulb, and the interesting perennial forms such as the Egyptian onion (*A. cepa* aggregatum), which has clusters of bulbils at the apex of the plant in place of flowers.

Pea

The garden pea is known as *Pisum sativum* but its origins are obscure. It has been cultivated since antiquity and presumably originated somewhere in the Middle East or Mediterranean region where related species still occur. I have mentioned previously that the pea is the most extensively grown British vegetable, and thus it clearly thrives in a cooler climate than that of its homeland. Nonetheless, most cultivars are not plants for the far north, although the selection that has taken place for larger pods and larger seeds has also included the development of some tolerance of cold. One feature that has remained constant is the plant's climbing or scrambling habit; it depends naturally very much on other vegetation around which it twines its leaf tendrils and, as a garden plant, dead birch or other twigs generally provide it with the best support. However, the most recent development in pea breeding has been in quite the reverse direction with the so-called leafless pea, in which the plant virtually comprises a mass of tendrils. The reasoning behind this apparently retrograde step is that, with less leaf, the plant can direct all its resources to pod production and that the masses of intertwining tendrils render a group of plants self-supporting. However, the evidence seems to be that these benefits are countered by the fact that, with less green tissue, the plant is unable to photosynthesise sufficient nutrient.

Beans

The beans grown in gardens belong to three quite separate species and two genera. The broad bean and the agricultural field bean,

Normal pea

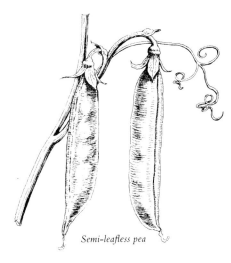

Semi-leafless pea

are both forms of the annual *Vicia faba*; the runner and French beans are derived from *Phaseolus multiflorus* and *P. vulgaris* respectively. *Vicia faba* is one of the most ancient of cultivated plants and, whenever beans are mentioned as either human or animal feed in historical writings, it is to this species that they belong. It almost certainly originated in the Mediterranean region, and a similar species occurs wild in North Africa. Nonetheless, many centuries of selection and hybridising have resulted in forms that are fully hardy in our climate. Indeed, among the recognisably distinct groups now cultivated, some, represented by the cultivar Aquadulce, are sufficiently hardy to survive outdoors over winter after late autumn sowing, and thus give the earliest crop in the following year. The seeds of broad beans vary widely in colour from deep purple to yellow-green.

The French, kidney or dwarf bean is a half-hardy annual, originating in South America and not suitable for the northerly parts of Britain. Like the broad bean, it is a free-standing plant but of much lower stature. It grows and matures quickly but is prone to wind bruising in exposed gardens. The trend among modern cultivars has been in favour of those which carry the beans well clear of the rest of the plant and which are therefore less likely to suffer from the rotting that arises from the contamination in rain-splashed soil.

The runner bean, or, as it is sometimes still called, the scarlet runner, was originally introduced to Europe from South America as an ornamental climber. It is a tall plant that scrambles through the low branches of trees in its native environment and, in gardens, must be supported by canes, wires or other means. Dwarf forms, possibly originating by hybridisation with the French bean, also exist. Although the original scarlet runner is the commonest form, others with white or scarlet-and-white flowers also occur, with

seeds ranging from black through the familiar black-and-purple to white. Purple podded types exist also. Unlike all other beans and peas grown in gardens, the runner bean is a perennial; the roots may be lifted in autumn, stored in the same way as dahlia tubers, and replanted to give a few very early pods in the following season.

Celery

Celery and its 'turnip-rooted' relative celeriac, are forms of *Apium graveolens*, a member of the Umbelliferae that occurs extensively throughout Europe and Asia, including Britain. It occurs naturally at river edges and in other marshy habitats, and the key to its successful cultivation lies in the provision of a copious water supply. It is commonly grown in well-manured trenches; the difference between trenched and non-trenched celery is described on p. 161.

Tomato

The tomato has much in common with its close relative the potato. Its wild ancestor is *Lycopersicum esculentum*, a native of South America, introduced to Europe in the middle of the sixteenth century. Like the runner bean, the tomato was first grown as an ornamental and, although the yellow-fruited cultivars (despite their generally better flavour) are little grown in gardens today, the original introductions were almost certainly of this colour and were called golden or love apples. The latter name, incidentally, persisted until well into the nineteenth century. The family Solanaceae, to which both tomatoes and potatoes belong, includes many of the most poisonous of plants, including the thorn apple (*Datura stramonium*) and deadly nightshade (*Atropa belladonna*) and, not surprisingly, the fruits of the tomato were for long regarded with great suspicion. Until the nineteenth century, garden catalogues listed them as ornamentals.

On p. 183 I discussed the difference between the determinate or bush tomatoes and the indeterminate or staked cultivars. The modern tomato produces a wide variety of types of fruit, from the very large Marmande and Beefsteak forms, through the more familiar round red-and-yellow fruited

cultivars, to tiny cherry tomatoes like Minibel, and the numerous elongated types familiarly illustrated on the cans of imported Italian 'plum tomatoes' but rarely grown successfully in Britain.

Potato

The potato is arguably the most important plant ever to have been introduced into Europe. The story of its arrival from South America, and of the most improbable part played in this by Sir Francis Drake and Sir Walter Raleigh, is familiar from most people's school days. It is certain, however, that the ancestor of the modern potato, *Solanum tuberosum*, together with other related tuber-forming species of the very large genus *Solanum*, had been cultivated for many years in South America before they were brought to Europe after the Spanish conquest in the late sixteenth century. Potato tubers are swollen stems, not roots, as evidenced by the fact that they produce chlorophyll and turn green on exposure to light. Potato cultivation is too well known to justify any repetition, but the possible use of seed in the future production of the crop is of interest.

Potato seeds are borne in green, tomato-like (but poisonous) berries; if these are sown they will produce new plants directly, just as wild potato seeds do. This could obviate the need for labour-intensive tuber production, costly storage and transportation, and circumvent the virus transmission that occurs so readily in tubers. However, potato seed is neither easy nor quick to germinate, the establishment of the seedlings is very chancy, and there remains the problem of crop uniformity. Potato clones are morphologically and physiologically uniform, characteristics exceedingly difficult to retain in the sexually-produced progeny of a plant that is genetically very complex. What is superficially an attractive alternative, therefore, is in practice bound up with many problems, and serves to indicate that cultivation has come a very long way from the simple copying of nature.

Cucurbits

Cucurbit is the group name given to cucumbers, marrows, pumpkins, squashes, gourds, melons and similar members of the family Cucurbitaceae. It is something of a wonder that they are grown successfully at all in temperate climates such as that of Britain, for all are either tropical or sub-tropical, and certainly none is fully hardy. They have all been cultivated for many centuries in warmer parts of the world and are climbing or scrambling annuals, securing themselves by means of stem tendrils. Their cultivation is interesting: cucumbers are usually grown as climbers; but marrows and pumpkins, because of the great weight of the fruit, are usually grown in their natural manner, trailing over the ground, where they occupy a very great deal of space in relation to the quantity of fruit produced. Melons are traditionally grown as climbers but require nets to support the ripening fruit.

Probably the hardiest of all the cucurbits are ridge cucumbers which are sufficiently hardy, and fast enough maturing, to produce abundant crops when grown outdoors during the summer. Indeed, so satisfactory are the modern cultivars that greenhouse space is much better given to other plants; there really is no reason why, over much of Britain, the Telegraph type of greenhouse cucumber should not become a thing of the past.

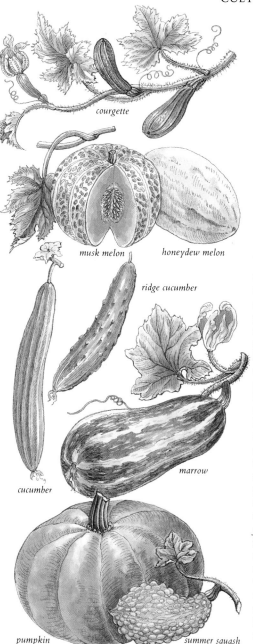

courgette

musk melon   honeydew melon

ridge cucumber

cucumber

marrow

pumpkin   summer squash

Dahlia

One of the largest of all plant families is the Compositae, often called the daisy family, and it has given us many of our most popular garden flowers, including asters, chrysanthemums and dahlias. The modern garden dahlia has its origins in several Mexican species, possibly including *Dahlia coccinea*, *D. pinnata* and *D. rosea*, although the plant was cultivated by the Aztecs long before the Spanish conquest and a wide range of hybrids was known even then. Nonetheless, the wild dahlia species tend to have single flowers (that is, flowers with a single row of ray florets) although almost all display the propensity to produce a proportion of doubles and semi-doubles among their progeny. Different wild species have contributed to different features of the modern garden plant; a plant known as *D. juarezii* (above), for instance, gave the spiky florets of the cactus dahlias. Much of the original appeal of the dahlia lay, incidentally, not in its flowers at all, but in its tubers which were thought a possible new food crop. The monstrous forms of the present-day giant show decoratives lead one to wish that breeding and selection had persisted rather more in a culinary direction.

Chrysanthemum

Akin to the genus *Dahlia* in the family
Compositae is *Chrysanthemum*, a much larger
group of over one hundred widely distributed
species. It is not known with certainty which
of these was the ancestor of the present day
garden or florist's chrysanthemum, although
the Chinese *C. sinense* or *C. indicum* are
strong candidates. The wild forms of these
plants are recognisably related to such
British species as *C. parthenium*, feverfew, in
their yellow disc and white ray florets. The
plant had been cultivated for centuries in the
Far East before it was introduced to Britain as
recently as the end of the eighteenth century.
Careful selection and hybridisation have
given rise to the range of modern cultivated
chrysanthemums, which vary almost as
widely in cultural requirements as they do in
form. Although there are annual types that
can easily be raised from seed and grown
outdoors, at the opposite end of the spectrum
lie the late-flowering greenhouse species,
propagated by cuttings and requiring a
complex regime of stopping and disbudding
to produce the enormous show blooms, more
like sculptured lamp-shades than anything
botanical.

Rose

To try to encapsulate the history of rose
cultivation, and the relationship between
wild and cultivated forms, in a few short
lines would be to insult a genus that spans a
range from the most exquisite of man's
embellishments upon nature's raw material to
some of the most hideous horticultural freaks
ever perpetrated. I shall not try, but shall
comment that the genus *Rosa* is noteworthy,
not only in displaying a very wide range in
habit from lax shrubs to majestic climbers,
but also in having contributed a particularly
wide range of species to cultivation. Table 19

Table 19   Some of the more important contributions of *Rosa* species to gardening*

Rosa species	Geographical origin	Growth habit
ecae	Afghanistan	Small, prickly shrub
foetida	Asia	Vigorous, erect shrub
hugonis	China	Vigorous, erect shrub
pimpinellifolia	Europe	Compact, very prickly shrub
sericea	Himalayas, W. China	Vigorous shrub, large thorns
xanthina	China	Angular shrub
gallica	Europe, S.W. Asia	Small, upright shrub
canina	Europe	Loose, straggly shrub
eglanteria	Europe	Loose, straggly shrub
glauca (= rubrifolia)	Europe	Upright shrub
carolinae	N. America	Small, suckering shrub
virginiana	N. America	Upright, bushy shrub
macrophylla	Himalayas	Upright, tall shrub
moyesii	W. China	Tall, vigorous shrub
multibracteata	China	Large, compact shrub
nutkana	N. America	Vigorous, upright shrub
rugosa	W. Asia, Japan	Vigorous, upright shrub
arvensis	Europe	Vigorous rambler
brunonii	Himalayas	Vigorous, very leafy climber
filipes	China	Vigorous climber
helenae	China	Vigorous climber
moschata	S. Europe	Climber
multiflora	E. Asia	Vigorous climber
sempervirens	S. Europe	Vigorous climber
setigera	N. America	Trailing shrub
sinowilsonii	China	Vigorous climber
soulieana	China	Vigorous shrub
wichuraiana	China	Dense shrub or rambler
chinensis	China	? not known in the wild
gigantea	Himalayas	Vigorous climber/shrub
banksiae	China	Climber
laevigata	China	Vigorous climber
bracteata	China	Climber
roxburghii	China	Tall shrub
stellata	N. America	Small, dense shrub

* *As many again are grown as species without having been hybridised.*

Flower form	Example of familiar clone	Important role in rose cultivation
Single, yellow	'Golden Chersonese'	
Single, yellow	'Lawrence Johnston'	A major source of yellow pigment for modern roses
Single, pale yellow	'Headleyensis'	
Single, creamy-pink	'Frühlingsgold'	Burnet Roses; many hybrids and forms
Single, (4-petals) white	'Hidcote Gold'	Several hybrids and forms with delicate, ferny foliage
Semi-double, yellow	'Canary Bird'	
Single, pink	'Jenny Duval'	An ancestor of the cabbage and moss roses, of the damasks, portland roses and probably several other hybrids
Single, pale pink		Probably an ancestor of the *alba* roses; formerly much used as a rootstock
Single, pink	'Lady Penzance'	Sweet Briars
Single, pink	'Carmenetta'	
Single, pink	'Double Pennsylvanian'	
Single, pink	'Rose d'Amour'	
Single, pink	'Doncasterii'	
Single, deep crimson	'Geranium'	
Single, lilac-pink	'Cerise Bouquet'	
Single, lilac-pink	'Cantab'	
Single, pink-red	'Blanc Double de Coubert'	Many other forms and hybrids
Single, white	'Splendens'	
Single, clustered, cream	'La Mortola'	
Single, clustered, cream	'Kiftsgate'	
Single, clustered, cream	'Lykkefund'	
Single, clustered, cream	'Paul's Himalayan Musk'	Hybrid musks and, indirectly, almost all modern roses
Single, clustered, cream		Hybrid musks and, indirectly, almost all modern roses
Single, clustered, white	'Felicite et Perpetue'	Probably many modern climbers
Single, clustered, pink	'Baltimore Belle'	Several other forms and hybrids
Single, clustered, white	'Wedding Day'	
Single, clustered, white	'Kew Rambler'	Most modern ramblers and many climbers
Single, white	'Albertine'	A large group of shrubs and climbers
?	'Cecile Brunner'	A parent of the Tea Roses
Single, white		A small group of partly hardy forms such as *lutea*
Single, white		
Single, white	'Silver Moon'	
Single, white	'Mermaid'	
Single, pink		A few forms, including at least one with double flowers
Single, pink-purple	'The Sacramento Rose'	

lists some of the more important contributions made by different wild species to modern cultivated types.

It may be common knowledge, but the fact bears repeating that virtually all rose species are thorny, an attribute that serves the dual purpose of providing them with protection from potential browsing and supporting them on and over other vegetation. This latter feature should not be forgotten and many, perhaps most, roses (apart from the modern bush types) will benefit from some form of support. The glory of the old shrub roses is always seen to its best advantage when they are given the support of a surrounding framework of lightweight rustic posts. And although the closer a rose is in breeding to a wild species, generally the shorter is its flowering season, this should serve as no deterrent to growing those less transmogrified by modern breeding. I have a bed of old shrub roses that is never without some bloom from late spring to very late autumn, simply from careful selection of types with overlapping flowering times.

Narcissi

Numerically, the genus *Narcissus* is the most important bulbous ornamental in British gardens. The majority of the millions of bulbs sold and planted each year belong to the group with elongated trumpet-like petals; they are generally known as daffodils, the name 'narcissus' being popularly restricted to the remainder. Garden daffodils are all hybrids, raised from several species, including the native British wild daffodil *Narcissus pseudonarcissus* and the Tenby daffodil *N. obvallaris*, but with important contributions from *N. cyclamineus, N. bicolor, N. poeticus, N. tazetta* and *N. jonquilla* among others. It should be remembered that most *Narcissus* species originate from the western Mediterranean region, and several are only

marginally hardy in Britain despite being dormant through the worst of our winter. Some, like the tiny *N. asturiensis* (above), although of southern distribution, are plants of the high mountains and much more resilient to British weather than might at first be thought. Moreover, some of the southern species that flower very early and succumb to the worst possible British weather are well worth growing in a cold greenhouse.

Tulip

Second only to narcissi in popularity among bulbs in Britain, tulips have waxed and waned in favour among gardeners; and although their export is now a major feature of the Dutch economy, no present day enthusiast can compare with the fanatical collectors of the early seventeenth century who paid kings' ransoms for single bulbs of choice hybrids. The genus *Tulipa* comprises around 100 species, distributed widely through Europe (*T. sylvestris* is fairly widely naturalised in parts of Britain), western and central Asia and North Africa, and the first centre of cultivation seems to have been in Turkey from where hybrids were brought to Europe in the late sixteenth century. These early hybrids were called *T. gesneriana* and from them have been derived most of the Darwin, double, paeony-flowered, lily-flowered and other modern garden forms. Much more rewarding, however, are the numerous very hardy and generally smaller *Tulipa* species, mostly from western Asia, now becoming increasingly available. By careful choice of species, it is possible to have a flowering season extending from early March to June; but it should be remembered that most are plants of rocky, mountainous regions, and a well-drained soil is a major key to their successful cultivation.

Tulipa turkestanica

Geranium and Pelargonium

The family Geraniaceae includes two important, distinct and unfortunately confused genera, *Geranium* and *Pelargonium*. They are most conveniently distinguished as hardy and half-hardy respectively in the British climate. *Pelargonium* is a predominantly African genus of around 250 species. The modern garden pelargoniums fall conveniently into the three groups of regal, zonal and ivy-leaved, all of which have a complex ancestry involving several African species. The zonal pelargoniums are the most familiar and it is these plants that are raised in their thousands every year by municipal parks and used for summer bedding, an activity that has been immeasurably simplified in recent years by the introduction of F_1 hybrid cultivars that can be raised annually from seed. Among the species to contribute to the modern zonals were *P. zonale* and the vividly red-flowered *P. inquinans*. The ivy-leaved or trailing pelargoniums are derived from *P. peltatum* and have changed much less in cultivation; only very recently has a cultivar been introduced that can be raised from seed annually. The most flamboyant flowers are those of the regal pelargoniums which are genetically the most complex of all; it is thought that at least eight African species have been involved in their development. Hence it should be remembered that, above all, they are flowers of a warm and sunny climate; as annuals in Britain, they need at least six months to reach their full flowering potential, which means sowing the seed in warmth in January or February. Being plants of a dry climate too, they are tolerant of considerable drought and, even in the hottest of British summers, can usually be relied on not to have wilted when a gardener returns from his two-week holiday.

The hardy genus *Geranium* is considerably larger than *Pelargonium* and more extensively distributed, occurring almost world-wide in temperate climates; there are about twenty native or naturalised species in Britain. Compared with pelargoniums, hybridising of geraniums has been minimal and the genus is a classic instance of the value of virtually unaltered species as garden plants. The habitats and life-forms vary widely among the group; most are fairly tall-growing perennials of open woodland and meadows (*G. pratense* and the beautiful, shade tolerant *G. phaeum*, for instance), whereas others, like *G. cinereum*, are creeping or cushion forming alpines. I believe that, with their variety of form, tolerance of a wide range of environmental conditions, rapid growth and generally very floriferous habit, geraniums are among the most rewarding and valuable of garden plants.

Iris

Few plant genera are as confusing to the gardener as *Iris*, for not only is it a big genus (around 200 species, although only five native or naturalised species in Britain), it is also one that spans a particularly wide range of form and cultivation. Although the following are not strict, botanical divisions, I find that these groupings are useful in summarising the ecology of the irises and in indicating their very different cultural requirements. First are the small, spring-flowering bulbous irises, mostly originating from the Mediterranean region and the Caucasus. They are generally grown in rock gardens or alpine beds and the blue *I. reticulata* and *I. histrioides* and the yellow *I. danfordiae* (right) are those seen most frequently. For some reason, most require planting at least 15 cm deep if they are to continue flowering perennially; otherwise they produce masses of daughter bulbs which must themselves grow for several years to attain flowering size.

The large bulbous irises are mostly derived from *I. xiphium* and *I. xiphiodes*, which occur principally in the Iberian peninsula. They grow well in a well-drained soil and full sun, and make excellent garden plants despite producing their foliage very early when it might be expected to be damaged by frost.

The bearded irises are rhizomatous, not bulbous. They are the most familiar garden irises and fall into three groups: the June flowering tall bearded; the April or May flowering dwarf bearded; and the intermediates which lie between the other two in flowering time and size. They have a most

complex ancestry involving *I. pallida*, *I. germanica*, *I. trojana* and *I. variegata*, mostly from south eastern Europe and western Asia. The so-called Californian hybrids are a group of dwarf rhizomatous irises that flower early, briefly, but gloriously, and merit wider attention as garden plants.

The waterside irises are a group of unrelated irises that are all plants of the water's edge or of mud, invaluable for bog or water gardens. The best known are the Japanese *I. kaempferi* and *I. laevigata*, the Siberian irises including *I. sibirica*, and the great yellow flag *I. pseudacorus*.

Iris unguicularis is a Mediterranean species, remarkably hardy and uniquely valuable for its flowering season which extends from late November until February. It requires a poor, free-draining soil and, as befits its origin, a baking in summer.

Finally, to the native British *I. foetidissima*, a plant commonly found on sea cliffs and with unremarkable purple-and-yellow flowers; but it has two major attributes in its vivid red seeds and its tolerance not only of dry soils but, unexpectedly, also of shade.

Clematis

There are two unusual features about *Clematis* as a genus of garden plants: first, that the range of species in existence is wholly unappreciated by most gardeners who grow only a small number of large-flowered hybrids, and one or two species; second, that its familiar climbing habit is not ubiquitous and can, in its most humble expressions, give plants that are better suited to the herbaceous border. I could, indeed, add a third feature, in that *Clematis* is almost unique in the family Ranunculaceae (the buttercup family) in producing woody tissues. In total, there are well over 300 species widely distributed throughout the north temperate zone, including many from China and the Himalayan region. The familiar large flowered hybrids are derived mostly from the European *C. viticella* and *C. integrifolia* and from the oriental *C. florida*, *C. lanuginosa* and *C. patens*. Among other excellent climbing species are *C. alpina* (below), *C. montana*, *C. texensis*, *C. orientalis* and *C. tangutica*. One of the best 'herbaceous' species is *C. heracleifolia* from China, while some of the Asian evergreen species, although too tender for garden culture in Britain, make splendid cool greenhouse plants.

Chapter Seven

Animal life and the garden

The garden, like any other ecological environment, is a more or less interdependent blend of representatives of all five of the kingdoms described on p. 78. The animal kingdom contains the greatest number of species, if not the greatest number of individuals; that distinction must belong to the Monerans for it has been estimated that bacteria alone account for over three quarters of all living organisms – a mere puddle of bacterial slime may contain more individuals than all of the animals in the world put together. Nonetheless, the numerical importance of animals should not be underestimated, for beetle species alone outnumber flowering plants (Table 20).

What are the characteristic features determining that an organism should be called animal rather than plant or something else? The next door neighbour's cat that scratches up your seed bed, and indeed the next door neighbour himself, are quite clearly not plants: they are not green, and they can walk. They are not green because they do not contain chlorophyll and do not photosynthesise; they obtain their food by

Table 20 Approximate numbers of species within the main groups of animals

Group	Approximate number of species		
Sponges	5,000		
Coelenterates (*jelly fish and allies*)	9,000		
Platyhelminths (*flatworms and tapeworms*)	13,000		
Nematodes (*eelworms, roundworms*)	13,000		
Annelids (*earthworms, ragworms, leeches*)	9,000		
Molluscs (*slugs, snails, shell-fish*)	35,000		
Arthropods	1,090,000		
[Insects	1,000,000]	The major groups of insects are:	
[Crustaceans (*crabs, lobsters, shrimps*)	30,000]	Dragonflies	5,000
[Arachnids (*spiders, mites, scorpions*)	50,000]	Crickets, locusts, grasshoppers	20,000
[Myriapods (*millipedes, centipedes*)	11,000]	Thrips	5,000
Echinoderms (*sea urchins, starfish*)	5,500	Aphids and allies	60,000
Cartilaginous fish (*sharks, rays*)	620	Beetles	350,000
Bony fish	20,000	Flies	70,000
Amphibians	2,750	Butterflies and moths	165,000
Reptiles	6,000	Wasps, ants, bees	110,000
Birds	8,600		
Mammals	4,500		

holozoic nutrition or, in everyday terms, by eating other organisms or the products of other organisms. The cat eats mice or fish (other organisms) and drinks milk (the product of other organisms). The next door neighbour may also drink milk and eat fish. He is unlikely to eat mice, but would probably not be averse to cows, sheep and pigs. Unlike the cat, he also appreciates fruit, vegetables, and, like the cow, sheep and pig in turn, the products of certain species of grass in the form of cereal grains, flour and flour products. From the simplest, microscopic forms of animal life to the highest stage of animal development, nutrients rather more complex than carbon dioxide, water and soil mineral elements are needed to sustain continued existence.

Movement is not exclusive to animals, for most protists and many monerans possess it also; indeed, many plants have limited powers of movement; the sunflower whose head always faces the sun clearly moves to a limited extent, but what plants *do* lack is the ability for directed self-sustaining movement from one location to another – locomotion. Hence, by widely accepted definition, John Wyndham's triffids could not have been plants. By and large, animals move because of their mode of feeding; unlike plants, whose food requirements can generally be satisfied without their needing to seek it out or hunt for it, most animals would very soon exhaust the food reserves that could be reached from a sedentary position. (There are exceptions; but usually, where an animal is at least partially stationary, it is because it lives in an environment where its food is brought to it.)

The final major difference between plants and animals concerns cell structure. It is the rigidity afforded by the substance lignin that has enabled an organism the size of a tree to arise. In fact, at a more fundamental level, almost every basic structural cell of a plant has a rigid outer wall, be it lignified or not. The typical animal cell is essentially a floppy little bag full of watery protoplasm; individual cells may contribute towards structural strength (skeletal cells, for instance) but they are the exceptions. The blue whale, far and away the largest animal that has ever lived, is nowhere nearly as large and heavy as a redwood tree; yet, lacking rigid cell walls, it requires the buoyancy of sea-water to exist at all.

Blue whales may seem as far from your garden as they do from mine; yet I make no apology for a brief diversion into the features of animals, for it usefully prefaces an understanding of the relative roles of plants and animals in the ecosystem that is your garden, and my suggestions as to how, if at all, you should respond to them.

I make no apology for adopting an anthropocentric approach and categorising garden animals into the beneficial and the pestilential. Gardening may be directed natural history and we may be able to explain the functions of the garden by reference to the events in a more natural ecological habitat, but it is no use pretending that we can garden in harmony with all forms of animal life. Just as plants in a garden need our attention to provide the food and water that cultivation has denied them, so they need assistance in combatting those animals in whose favour we may have tipped the balance of the environment.

Beneficial animal life in the garden

In Chapter 1, I discussed the topic of nutrient cycling and the food web of immensely complex interrelationships between organisms. Animals play an especially important role in such interactions through their ability for long range movement, and through their relatively large size and substantial nutritional requirements. (It would take a great many bacteria or fungi a very long time indeed to achieve the disintegration of as much plant material as a cow's stomach manages in a few hours.) But food webs are only one of the important areas where plants and animals interact with consequences for garden ecology.

Pollination

The natural variation in flower form and structure is the result of evolutionary adaptations by plants to facilitate pollination by wind and animals of different types. Although we think of honey bees as the animals most importantly associated with the transfer of pollen, they are merely the most conspicuous. Many other species of bee, fly, butterfly and moth are typical pollinators of temperate climate plants; and almost every group of the animal kingdom has representatives that fulfil this role with one or another plant species and, in the tropics especially, there are plants dependent on slugs and snails, on birds, bats and most types of animal in between.

It must be admitted that there are relatively few instances where pollination is essential for a garden to be a success. By and large, a gardener is concerned more with the production of flowers or vegetative parts than with the production of seeds; the relatively few exceptions are when the seeds themselves, or the fruit that they contain, are edible, or when they add to a plant's ornamental appeal. Even then, some such

The most familiar and important role that animals play in gardens is as carriers of pollen for the fertilisation of flowers.

plants are wind-pollinated or set seed apomictically (without fertilisation at all). Moreover, even where pollinating animals are required (and in British gardens it is normally insects with which we are concerned), there is relatively little that can be done to encourage them if there is not already a good population in the neighbourhood. The growing of some types of plant may be advantageous in encouraging insects to visit a garden in the hope that they will then switch their attention to fruit trees or bushes, but more often than not, insects will find their way into any garden containing flowers; it is the problem of keeping them there long enough to fulfil their pollinating role in an exposed and windy site that most commonly creates problems. I refer you again, therefore, to the importance of establishing shelter (or not removing existing shelter), that I discussed in Chapter 2.

Seed and fruit dispersal

Robin

Wren

Chaffinch

Blackbird

House Sparrow

Great tit

Blue tit

Greenfinch

Norman Arlott.

In gardens, as elsewhere, birds are the most important among those animals that aid seed dispersal. Partly this is because they can move rapidly over long distances, but also because many seeds have outer coats able to resist birds' digestive enzymes. Thus, while the siting of seeds within the fruit on which birds feed ensures that they will be picked up, the protective coat ensures that they will be delivered safely. Indeed, in some instances, the digestive process has a positive benefit in eroding just sufficient of the seed coat to facilitate germination.

From what I have said about competition between plants, and about the need for individuals of a species to become dispersed from their fellows, it is clear that any animal activity that aids this is of positive benefit to the plant species as a whole. Nonetheless, among the features of gardens setting them apart from natural plant communities is that a gardener's wishes dictate the positions in which the plants grow. Anything disturbing his wishes and causing plants to grow elsewhere does not usually meet with his approval unless it falls into those few situations where he wants a plant to spread more or less randomly over a given area.

The occasional tree or shrub seedlings that appear uninvited among other garden plants are almost always the result of their seeds having being carried by birds. Hence my reason for categorising this a beneficial activity, for any potentially interesting tree or shrub that arrives, unheralded and above all *free*, is surely to be welcomed.

Bullfinch

Starling

Dunnock

Song Thrush

Table 21   Birds at bird tables – the beneficial and/or harmful roles played by some common garden birds*

Species	Principal food	Significance for gardeners
Wren	Insects, especially caterpillars, spiders	Positively beneficial
Dunnock	Insects, especially caterpillars, spiders, some seeds in winter	Positively beneficial; seeds are usually weeds
Robin	Insects, worms, some small fruit and seeds	On balance, beneficial
Blackbird	Insects, worms, snails, fruit	On balance, usually beneficial but can be very troublesome to ripening fruit and can deplete ornamental berries in hard winters
Song thrush	Insects, worms, snails, fruit	On balance, usually beneficial but can be very troublesome to ripening fruit and can deplete ornamental berries in hard winters
Blue tit	Insects, fruit, buds	Marginal; diet of large numbers of aphids and other pests must be counterbalanced by the damage to ornamental flower buds in early spring
Great tit	Insects, fruit, seeds	On balance, errs more than the blue tit on the beneficial side
Starling	Insects, some fruit	On balance beneficial; although they may cause some damage to fruit, their activities in depleting numbers of soil inhabiting insect pests are valuable, especially on lawns
House sparrow	Insects, seeds	Pest; insect eating activities are more than offset by the wanton destruction of buds on fruit trees and bushes and by the apparently pointless destruction of crocus flowers and seedling plants
Chaffinch	Insects, seeds	Positively beneficial; seeds are usually weeds
Greenfinch	Seeds, insects	Positively beneficial; seeds are usually weeds
Bullfinch	Seeds, buds, fruit, insects	Pest; insect eating activities are more than offset by the wholesale destruction of buds on fruit trees and bushes

* in categorising birds in this way, I must stress that all enjoy legal protection in Britain and any counter measures taken must no cause them harm in any way

Predatory and parasitic activity

Among the countless individual interactions between the animals in a food web will be some involving those that we define as pests. Thus, certain animals are seen as positively beneficial simply because their diet includes pest species.

If all the offspring of a single aphid on a broad bean plant early in the summer developed to maturity and they, and then their offspring in turn, all gave birth to the maximum number of young, your garden could be host to around one million tonnes of aphids within about two months. Fortunately, this prodigious reproductive potential is never fully realised, for the aphid population will itself be regulated by disease and by natural predators. Ladybird larvae may eat up to five hundred aphids each during the space of three weeks. Each hover-fly larva is fully capable of dining on a thousand or more aphids during the course of their development, sometimes at the rate of one every minute. And, of course, many familiar garden birds such as blue tits include aphids and other insect pests in their diet.

Aphids are the most familiar and in many ways the most serious of insect pests. Their sap-sucking activities weaken plants while their role as carriers of viruses adds another dimension to the importance of aphid control.

Some animals regulate the populations of other species by parasitising them, rather than eating them whole. The females of many different types of fly lay their eggs within the bodies of other insects which die when the eggs hatch and the larvae emerge. Never was the importance of this seen so dramatically by gardeners as during the population explosions of large white butterfly caterpillars (*Pieris brassicae*) during the 1950s. As the caterpillar populations on brassicas in gardens and allotments reached a peak, increasing numbers of them were seen to be parasitised by the larvae of ichneumon flies which, together with bacterial and viral diseases, reduced the numbers of butterflies to normal.

Non-beneficial animal life in the garden

As a weed has been called a plant growing in the wrong place, so a pest is an animal living and feeding counter to our endeavours. In almost all pest species it is their feeding that causes damage to garden plants (although a few incidental causes of damage, such as the scratching of bark or seed-bed by a domestic cat, can be at least as distressing). Thus, whenever we take action against a pest, even if we don't actually kill it but simply protect our plants from its advances, we are denying some organism its food source and this will have one of two consequences. Either the organism will die from starvation when, of course, its own species will be threatened, as will those others that in turn depend upon it for food, shelter, pollination or other benefit. Or the pest organism will

Table 22 The most frequently misidentified or unjustifiably maligned garden animals

Earthworms Condemned for fouling lawn turf with casts, yet only two of the twenty five British species are responsible and their roles in aiding organic matter breakdown and improving soil aeration and drainage more than compensate for any damage that they cause in this way

Centipedes Similar to harmful millipedes, but with fewer legs; carnivorous, with many garden pests in their diet, unlike millipedes which are vegetarian and often damaging to plant life

Spiders The archetypal creepy-crawly; eight legs differentiates from insects, which have six. Carnivorous, with a diet including flies, caterpillars and many garden pests

Wasps Disliked for the sting and damage to fruit late in the summer, but for much of the year, play a beneficial role in feeding on other insects, including pest species

Hover flies Correctly called syrphids; mistaken for wasps and condemned with them, but positively beneficial – their larvae are among the most important predators of aphids

Harvestmen Condemned in common with most long-legged creatures; confused with winged crane flies (the adults of leather jackets) but possess eight legs, are related to spiders and are similarly beneficial

Frogs and toads Toads in particular have a long history of being disliked by people and were once associated with witchcraft but both toads and frogs feed on insects and

Spider

Hedgehog

Toad

Cat

other invertebrates and play an important part in keeping pest populations in check

Slow worms Legless lizards, damned for their superficial resemblance to snakes but important predators of insects and other garden pests

Collared doves A recent arrival among British birds but now extremely common in many areas. Undoubtedly causes some damage to crops but nothing like as serious a pest as its larger relative, the wood pigeon

Shrews Superficially similar to mice, voles and other rodents but quite unrelated, insectivorous and responsible for eating vast quantities of garden pests

Hedgehog Insectivorous. Though rarely condemned, suffers as so commonly struck by cars on country roads at night

Red squirrels Condemned with its introduced relative, the grey squirrel, as 'tree rats' but now rare in England and any damage caused to trees should be overlooked in the cause of conserving one of the most beautiful British mammals

Cats An emotive subject, but, on balance, a well-trained cat more than counterbalances its scratching activities with keeping rodent populations in check. An untrained cat that is not naturally a mouser can be a liability nonetheless

The most troublesome garden pests are not necessarily the most obvious. Ants can cause serious problems through their activities in undermining plants. Increasing the organic matter content of the soil may aid control. .

move elsewhere (remember the advantage to an animal of locomotion), cause problems in someone else's garden, or even, after prolonged deprivation of its normal food source, turn its attention to some other plant species, perhaps one even more important as a crop.

A 'pest' is any animal that makes its presence apparent to the gardener in a particularly destructive fashion. But there are many forms of animal life that seem to fall into a sort of neutral category, that seem to do neither harm nor good. If a gardener adopts the maxim of treating as beneficial any form of animal life for which, in his garden, he knows of no clearly defined disadvantageous role, he will be erring on the right side. I have used the phrase 'in his garden' advisedly for there is a rather important distinction to be made between the general and the local roles. I can see neither merit nor moral justification in attempting to drive out an animal species that may cause problems elsewhere, if its presence in your garden is immaterial. Thus, unless you grow brassicas, the presence of large white butterflies in your garden is certainly no cause for insecticidal activity. I would, indeed, extend the maxim further and say that unless and until you are quite certain that any particular organism is causing significant and intolerable damage should you take action. I appreciate that reaching this state of being 'quite certain' and of knowing how much damage is 'intolerable' may not be easy, and in advocating this trend away from horticultural hypochondria, I will use a few examples to amplify my point.

Most types of garden plant will, at some time during the summer, support a population of one or more of the five hundred British species of aphid. Many of them are capable of weakening plants quite dramatically through their sap-sucking activities, and many are also important as vectors of plant viruses. Does this mean, therefore, that the first sign of aphids on a garden plant necessitates remedial action. No, it does not, for several reasons. The aphids that you see may not have the reproductive capacity to build up to significant and damaging levels on the plant concerned, or at least to do so before the plant's growing season is at an end; they may not be virus-carrying species; and their numbers may be kept in check by one of the numerous predators on aphids. I could probably count on one, or at most two hands, the garden plants that will need protection from aphids to give of their best in most seasons; roses and broad beans would be top of the list although I know also that, in my garden, the aquilegias will suffer in four years out of five, and,

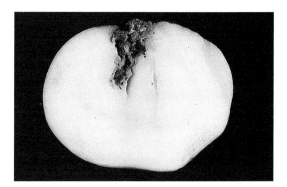

The codling moth larva is one of the commonest pests of apples but extremely difficult to control, even when its biology and method of entering the fruit are understood. Spray timing is so critical as to be almost impracticable.

whilst my mahonias usually escape, neighbours' plants tend to be infested. Possibly my mahonias have some inherent resistance to attack; more probably they occupy a niche in my garden that prevents aphids from building up – they are further from the house than my neighbours' bushes so perhaps blue tits are more willing to feed on them.

For a further example, consider a single insect species, the codling moth (*Cydia pomonella*), one of the commonest and most widely distributed pests of apples and one on which the commercial fruit-growing industry expends vast sums in chemical control. Most apple trees, in most gardens, will experience attacks by this insect every year. The female moths lay their eggs singly during June on the young fruits and on the leaves close to the fruits. About two weeks later, the young caterpillars emerge and immediately and insidiously burrow into the fruit, usually by the most inconspicuous route close to the 'eye'. Once inside, they feed contentedly for a few weeks before emerging to spin cocoons on the bark in which to pupate. Chemical spraying must be timed to coincide with the very brief period between the emergence of the caterpillars and their entry into the fruit; it must also be extremely thorough, as the eggs are widely dispersed. On large trees, this degree of thoroughness is quite impossible to achieve. The approach recommended in much of the older gardening literature was to tie sacking around the trunk of the tree, with the object of inducing the caterpillars to form their cocoons within it. The material was then removed entire and the caterpillars destroyed. Unfortunately, a little ecological thinking would have revealed that this was of somewhat improbable benefit. Caterpillars forming cocoons will already have damaged the current season's crop of apples and the impact on next year's population of female moths will be minimal, considering how common is the insect and the ease with which adults could fly in from neighbouring gardens.

I conclude, therefore, that control of codling moths is only worth attempting on young trees of manageable size (especially on the modern dwarfing rootstocks). I decided long ago that spraying my apple trees, ranging in age from fifty to one hundred years and in height from fifteen to thirty feet, was a futile pastime; we always have ample apples to eat, I have plenty of organic matter in the infested fruit to recycle through my compost, and I have one less reason for employing a chemical pesticide.

A third example will illustrate one more facet of the dilemma facing

Wasps are pests for a few weeks of the year only as they feed their young on other, often harmful insects. Even so, the damage they cause may be enlarged by other pests and diseases.

Although many fly larvae may damage plants directly the adult insects are merely attracted by the sweetness released when tissues have been damaged by other causes.

the gardener when he considers whether or not to take action against an animal. There are few more annoying aspects of late summer in Britain than the attention accorded to people and picnics by one or other species of *Paravespula*, the wasps. Understandably, they are swatted, chased with rolled newspapers and generally find themselves accorded attention commensurate with the status of major garden pests. All of this I can appreciate. There comes a point, however, when man versus wasp takes on a more global dimension and individual insects are followed assiduously to their homes in hollow walls, old tree trunks and the like, there to have their colonies destroyed with carbaryl or other chemical insecticide. I find this hard to condone (other than where the nest is very close to the house or where children or persons with severe allergic reactions may be stung), partly because the degree of damage caused by wasps to garden plants is relatively small; they will damage fruit to some extent, but usually this is merely an extension of wounds caused by birds or other agencies. More importantly, wasps are positively beneficial for much of the year when they prey on caterpillars and other insects, including many that are undeniably pest species.

Thus far, I have selected examples where the pest status of a particular animal is arguable or where its control is a fairly futile objective. While I can think of no species that merits a 'shoot on sight' approach, there are a few animals that may be expected to be seriously troublesome to anyone attempting to grow plants, and against which some general and continuing protection is advisable. In this small category I would place moles, grey squirrels, woodlice, millepedes and slugs, although even here only a minority of the British species of most of these animal groups actually causes problems to gardeners.

A particularly helpful attribute for any gardener is to be able to recognise as beneficial or pest those animals that pass through two or more stages in their life cycles. Two of the examples referred to already display extremes in this respect, the adult codling moth being most unlikely ever to be seen, let alone appreciated for what it is, whereas the large white butterfly will be known universally as the parent of cabbage caterpillars. Less widely appreciated are craneflies (or daddy-long-legs) as the adults of the soil-inhabiting leatherjacket, click beetles as the adults of wireworms, and some of the most attractive moths such as the large yellow underwing as adults of root-eating cutworm caterpillars. Sometimes, the relationship operates in reverse and, with crickets and others, it is the adult rather than the immature form that is the more serious pest. In yet others, adults and larvae both cause damage, either of different types as with flea beetles, or similarly as in the classic case of the Colorado potato beetle (*Leptinotarsa decemlineata*).

Sometimes the relationship between separate stages of the life cycle is confused because different types of plant are attacked at different times. Many species of aphid and adelgid migrate from one type of plant to

Both adult and immature woodlice damage plants; more or less in proportion to their size.

Adult crane flies are harmless to plants but their leatherjacket larvae cause great damage.

Both larvae and mature colorado beetles (above) devour the foliage of potatoes and related plants. Only by the strictest quarantine are these pests excluded from the British Isles as the consequences for the British potato crop could be catastrophic if they were to become established here. By contrast, crickets and grasshoppers (left) are relatively harmless for although they are primarily vegetarian, grass forms the main part of their diet.

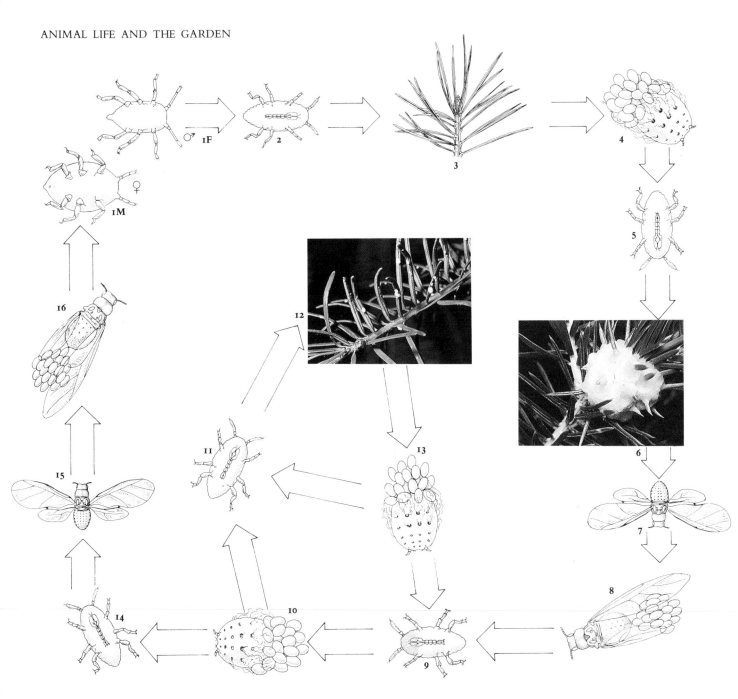

another and, as with fungal rust diseases, it may be possible to protect garden plants by taking special care to eradicate certain of the weed species on which these animals feed (Diag. p. 210).

Overall, therefore, I would promote the notion that a little learning is better than no learning at all. It is not unreasonable to suggest that you learn to distinguish between a probably harmful millipede and a definitely beneficial centipede, between an eight-legged spider and a six-legged insect, between an eight-legged spider and an eight-legged mite, and even between a harmful red spider mite and a beneficial orobatid mite. And learn sufficient of animal biology to appreciate the relationship between larva and adult of at least the more conspicuous and important among garden pests.

The control of pests

*Life cycle of a conifer adelgid (*Adelges laricis)
have several distinct phases or morphs.
1F Female Sexuales *in summer*
1M Male Sexuales *in summer.*
2 Young Fundatrix *on spruce shoot in autumn.*
3 Developing Fundatrix *near spruce bud in spring.*
4 Adult Fundatrix *in spring.*
5 Young Gallicola *on spruce.*
6 Open spruce gall in summer.
7 Gallicola flying from spruce to larch trees in summer.
8 Gallicola on larch needle with egg cluster.
9 Young Sistens *on twigs and long shoots of larch through the autumn and winter.*
10 Adult Sistens *with eggs on larch twigs in spring.*
11 Young Progradiens Aptera *on larch needles in spring and summer.*
12 Progradiens *colonies on dwarf shoots of larch in spring and summer.*
13 Adult Progradiens Aptera *on larch needles.*
14 Young Progradiens Alata *on larch needles.*
15 Sexuparae *or* Progradiens Alata *flying from larch to spruce trees in early summer.*
16 Sexupara *with eggs on young spruce needles.*
(Based on an original drawing by C. I. Carter.)

Books, big books, have been written on the subject of pest control in gardens and I don't propose to repeat all their advice here. Nonetheless, it is appropriate to consider the three principle approaches to the subject in relation to their impact on garden ecology as a whole, bearing in mind always that any action you take will inevitably have repercussions beyond the obvious.

First, think carefully about the need for any artificial control; will nature manage on her own? Second, if action is necessary, can you use a non-chemical or cultural approach? Third, if you do require a chemical aid, how can you be sure that the hoverflies and ladybirds won't be swept away by it too, or that the blue tits won't be poisoned? Experience has a great deal to do with the first question, for only by experience can you tell whether particular plants in your garden are likely to suffer significantly and regularly from any particular pest.

Once you have decided, however, that natural pest-regulating mechanisms cannot cope with the unnatural environment that is your garden, is the time to see how much you can help them. I have always advocated that a clean and tidy garden is likely to be a reasonably healthy one, for nothing so harbours pests such as woodlice and millipedes as much as piles of garden debris, be it in the form of crop remains, old seed-boxes or the piles of rubbish that so often pass for compost heaps. Grow your plants well feed, water and nourish them in the ways that best suit them, rotate your vegetable crops, and you will have played a significant card in the game of pest avoidance, if not quite control.

Theory suggests that many species of pest will die away in the soil if

The holes betray the activities of elm bark beetles, carriers of the elm disease fungus. Theoretically, control of the insect should aid control of the disease – but is difficult.

Simple, but very effective, are sticky grease band traps on fruit trees which depend on knowing that the flightless females of winter moths and related pests must crawl up the trunk to lay their eggs.

they are deprived of their host plants for a number of years, the actual number varying with the particular pest (Table 23). This is often true in commercial practice, where the rotated plots may be a kilometre apart. In a garden, however, this year's carrots will be only a few strides from the plot that bore them in previous years, and, if you can stride this distance, how much more easily can a carrot-fly fly it?

The first step from the more or less total reliance on sound husbandry and nature is to use a hose-pipe to combat aphids, half-sunken saucers full of beer to trap slugs, and soapy water sprays to wash away caterpillars; there is a bevy of such home-spun remedies ranging from the evidently logical, if obscure, to the blatantly bizarre. Some seem to work for some people; others seem to work for no-one and I often wonder how they originated. Nonetheless, there is no denying that soapy water, will wash aphids, caterpillars and other pests from plants. They may, in time, recover and simply crawl up again and it may be impossible to use sufficient force to knock off the insects without damaging the plants. Nipping out the young shoot tips of broad beans will remove potential feeding sites for aphids and, while such procedures usually fall short of chemical controls in effectiveness, my advice is, if they work for you, then use them by all means; but don't expect too much.

The most contentious aspect of pest control relates to the use of chemical pesticides. All chemicals available to gardeners in this country must meet standards of safety when used in the ways, at the times and in the doses prescribed by the manufacturers. But all chemicals, even the salt from your dining table, can be harmful to plants, animals and people in some circumstances. Hence the importance, which cannot be over-stressed, of following *to the letter* the manufacturers' directions: read them once, then *read them again* before you open the bottle.

To over-simplify a complex subject, I would suggest that there are two main ways in which the acceptability of pesticides for garden use should be judged. I ask the questions: 'Does the product have specific or broad-spectrum toxicity?' and 'How long does the chemical persist in the environment in toxic form?'. Unfortunately, reading the label on the bottle will not give you the full answers to either. Broad-spectrum toxicity means that the product is likely, at the recommended rates, to kill a wide range of different creatures. Most garden insecticides are toxic to most insects, although to varying degrees. The notable exception is pirimicarb, which is effective only against aphids and is invaluable therefore in leaving pollinating insects unharmed. But, given a product with wide-ranging activity and a desire not to harm beneficial insects, I suggest that you spray only late in the evening when most pollinating activity has finished.

Degree of persistence in the environment is also not easy to discover although the guidelines given on product labels regarding the minimum period that must elapse after treatment and before fruit and vegetable crops may be eaten is a good indication. In general, systemic insecticides which are absorbed into the plant's sap will give longer-lasting protection but are also, inevitably, present for longer periods. Here again, I can do no more than indicate my own maxim, which is not to use systemic chemicals on edible crops, and only to use them at all if you have failed totally to control a serious problem on an ornamental with

Table 23 Persistence in the soil of some common garden pests

Pest	Soil inhabiting phases of life cycle	Length of rotation likely to bring about significant diminution of pest*
Eelworms (some species)	All (at certain times)	From 2–3 years (stem eelworm) to 7–8 years (pea cyst eelworm)
Slugs (some species)	All	Unlikely to have any effect as pest is ubiquitous
Millipedes	All	Unlikely to have any effect as pest is ubiquitous
Woodlice	All	Unlikely to have any effect as pest is ubiquitous
Cabbage root fly	Eggs, larvae (in roots) and pupae	3 years
Carrot fly	Eggs, larvae (in roots) and pupae	3 years
Narcissus fly	Eggs, larvae (in bulbs) and pupae	3 years
Crane flies	Eggs, larvae (leather jackets)	4 years
Click beetles	Eggs, larvae (wireworms) and pupae	4 years
Raspberry beetle	Pupae and adults	Rotation impracticable but rest the soil from raspberries for 2 years
Chafers	Eggs, larvae and pupae	2–6 years, depending on species
Vine weevils	Eggs, larvae and pupae	2 years
Winter moths	Pupae	Rotation impracticable (fruit tree pest)
Noctuid moths	Larvae (cutworms) and pupae	2 years

* These figures are based essentially on the longevity of the pests in the soil and will, in practice, be affected by such factors as the presence of wild plants on which persistence can occur, the thoroughness of general cultivations and the seasonal weather conditions

a non-systemic or contact product. Other than for very special problems, moreover, my choice of contact chemicals is largely restricted to materials that are more or less natural plant products like derris and pyrethrum, or the modern artificial versions of them such as permethrin, resmethrin and bioresmethrin. I prefer not to add insecticides to the soil itself but, when it is necessary, I prefer bromophos to gamma-HCH. In Table 24, I have listed what I feel are among the more environmentally acceptable and unacceptable garden pesticides and their uses.

Finally, a mention of what must, in theory, be one of the most attractive propositions for the ecology-conscious gardener. It is possible to adapt natural pest regulation processes to your own use, with what are generally called biological control techniques. The two best known examples are the use of *Phytoseiulus persimilis*, a South American predatory species of mite to feed on red spider mites, and of a parasitic wasp-like insect *Encarsia formosa* to control whiteflies. Both techniques are appealing because the pests concerned are otherwise very difficult to control, but they are restricted to greenhouse use. Purchase a small 'starter culture' of the predator or parasite, introduce it into your greenhouse and let nature take its course. Once the course has fully run and most pests have been eliminated, the predator and parasite will themselves die out. They will not control other types of pest, and you must not use a chemical insecticide for this purpose because of the damage that will ensue to the predators and parasites. The technique, therefore, is much more valuable in the large monocultures of commercial greenhouses than in small garden buildings.

There is now available a 'biological spray' comprising a culture of bacteria for use in controlling caterpillars outdoors in the garden, and this seems comparable in effectiveness with at least some of the chemical alternatives.

I began this chapter with reference to those archetypal animals, your next door neighbour and his cat, and it is to them that I now return. In the face of all the forms of animal life that may plague your gardening endeavours, spare a special thought for those instances where your pest is another man's pet. There are times when we all feel frustrations at the activities of other people's cats or, worse, dogs in our gardens. Remember, nonetheless, that the best means of control more often arises from an agreeable exchange of views with the neighbour than from a campaign of vindictiveness against his animal.

Table 24 Nature, mode of action, mammalian toxicity and persistence of some commonly available garden pesticides

Pesticide	Chemical type	Mode of action	Used against	Acute LD^{50}* for rats (mg/kg body weight)	Persistence in environment**
Aluminium ammonium sulphate	Inorganic metal salt	Repellent	Birds	?	Medium
Bioresmethrin	Synthetic pyrethroid	Contact	Insects	7070–8000	Short
Borax	Inorganic metal salt	Cumulative	Ants	4500–6000	Medium
Bromophos	Organophosphorus	Contact/residual	Insects	3750–8000	Medium
Mercurous chloride	Inorganic metal salt	Cumulative	Insects	210	Long
Carbaryl	Carbamate	Contact/residual	Insects	850	Medium
Chlordane	Organochlorine	Contact/residual	Insects	457–590	Medium
Coumatetralyl	Naphthyl-coumarin complex	Anticoagulant	Rodents	0.3 [1]	Medium
Diazinon	Organophosphorus	Contact/residual	Insects	300–400	Long
Difenacoum	Naphthyl-coumarin complex	Anticoagulant	Rodents	1.8	Long
Dimethoate	Organophosphorus	Systemic/contact	Insects & mites	500–600	Short
Fenitrothion	Organophosphorus	Contact/residual	Insects	800	Short
HCH	Organochlorine	Contact/residual	Insects	88–270	Medium
Malathion	Organophosphorus	Contact	Insects & mites	2800	Short
Metaldehyde	Aldehyde polymer	Anaesthetic	Molluscs	600–1000 [2]	Medium
Methiocarb	Carbamate	Cumulative/repellent	Insects & mites	100	Medium
Permethrin	Synthetic pyrethroid	Contact	Insects	430–4000 [3]	Short
Phoxim	Organophosphorus	Contact/residual	Insects	1976–2170	Short
Pirimicarb	Carbamate	Contact/part systemic	Aphids	147	Short
Pirimiphos-methyl	Organophosphorus	Contact/part systemic	Insects & mites	2050	Short
Pyrethrum (pyrethrins)	Natural plant product	Contact	Insects	589–900 [4]	Short
Resmethrin	Synthetic pyrethroid	Contact	Insects	>2500	Short
Rotenone (derris)	Natural plant product	Contact	Insects	132–1500 [5]	Short
Tar Oil(s)	Hydrocarbon mixture	Contact/residual	Insects & mites	?	Long
Trichlorphon	Organophosphorus	Contact/residual	Insects	560–630	Medium

* the dose required to kill 50% of a population of test animals
** some of these estimates are subjective as detailed data are often obscure

[1] sub-chronic LD^{50} over 5 days
[2] dogs
[3] varies with age and sex of animal and precise chemistry of chemical highly toxic to fish
[5] highly toxic to fish and pigs

Chapter Eight

Ecology in garden history and garden style

Look at a plan of the gardens of Pompeii, of ancient Egypt, of any Mediaeval monastery or, indeed, of almost any other garden built over a period of many, many centuries and you will see some of the features that I have discussed in this book. Almost without exception, they are gardens of the monoculture and the straight line. It is a little understood and seldom remarked upon feature of early garden development that our ancestors did not attempt to mimic nature. When our ancient tribesman first decided to plant corn close to his village, he chose not to plant it in mixture with the other plants that accompanied it naturally. He also, to a greater or lesser extent, chose to confine its spread with borders of other plants, or with rocks or similar materials. From such beginnings, gardening history began, and innumerable trends and influences mark its progress to the suburban home garden of today. These trends varied in different parts of the world and they have been thoroughly and extensively researched and chronicled. My purpose here, however, is solely to analyse garden development in relation to ecological principles, and I shall confine myself geographically to the British garden.

Before I begin, however, one important point must be made. Until very recently, almost until a hundred years ago, the ornamental garden in Britain, with one important exception, was the grand garden. The poor and the ordinary man by and large were too busy with the task of living to cultivate an interest in ornamentals. If they did cultivate, in a literal sense, it was in a home vegetable plot, little different from those to be found among peasant communities in many parts of the world today. In Britain, if this plot survives at all, it is in the crofter's cabbage garden or, in modified form, I suppose, the allotment.

The one very important exception to which I have referred is the type of garden thought of by many as *the* English garden, the cottage garden. It has stood proudly outside the mainstream of garden development and it remained largely unchanged for hundreds of years, an essentially rural habitat for flowers, vegetables, fruit, pigs, chickens and people, co-habiting in a complete biological hotch-potch. Always small, sometimes minute, the cottage garden surrounded the homes of the rural lower classes and has served, incidentally, like a people's gene bank, as the refuge for many old plant cultivars and varieties. Ecologically, the cottage garden almost defies analysis although the cottagers themselves were more intimately bound up with the ecosystem than have been the owners of almost any other type of garden, before or since. They knew

The peasant garden no longer survives in Britain but many can still be seen in southern Europe.

Many modern gardens show cottage garden influence in their mixing of fruit, flowers and vegetables.

nothing of fertilisers and pest control and little of weeding, but a great deal of competition, invasion and succession, of self-seeding, of the merits of native plants and of complex communities. The cottage garden has endured little in its original form, although it has contributed one style to imitate in the modern home garden. It has endured most, however, in the inspiration that it gave to Robinson, Jekyll and their followers at the end of the nineteenth century, to which I shall refer again.

Ornamental growing among ordinary people was unusual in that it was also an urban phenomenon. The beautiful laced pinks and auriculas of the Paisley and Lancashire weavers moreover bear testimony to localised urban interest in particular groups of garden flowers. The gardens in which they were grown were minute, even if the plants themselves were princely.

And thus to the main thrust of British garden development. The earliest ornamental gardens of any significance in Britain were those of the Romans and although none survives, their outlines may be seen at several villa sites, as well as at the great Romano-British palace of Fishbourne. In their plants, they almost certainly reflected, insofar as climate permitted, the gardens of the continental Roman Empire, including, of course, Rome itself and Pompeii, where they have been well preserved. Their formal, generally simple, geometric patterns with paths, courtyards, gravel and a blend of plants in beds and in containers can trace an ancestry back through Egypt, Greece, Babylon (remember the hanging gardens) and other early civilisations. The plants were a blend of the herbal, medicinal, edible, symbolic and ornamental, and the gardens were unmistakeably artificial arrangements. From such beginnings, almost to the present, the guiding principle of gardening was to go against nature rather than with it. It seems that gardens merely reflected man's aggressive desire to subjugate both his fellow men and the environment in which they lived. Only by confining his trees, shrubs and other plants, by planting them in lines, in square blocks, in pots, pits, tubs and boxes could he demonstrate his supremacy over nature and the world in general.

After the departure of the Romans in the fifth century A.D., gardening in Britain was left in unknown hands until it surfaced over five hundred years later in the monasteries and, to some extent, in the universities. As befits a monastic existence, there was little other than the functional about these gardens, which cultivated herbs, vegetables, fruits and other useful plants, although the archetypal Mediaeval monastery garden had an incidental aesthetic charm in its arbours, courtyards, quadrangles and raised beds. Several present-day abbeys and monasteries retain, or have resurrected, examples of this type of horticulture.

From an ecological standpoint, we can telescope the next two hundred years of British gardening, during which the ornamental garden became an important adjunct to the stately home and the palace. While influences on gardening style came from many quarters, the relationship between planted and unplanted areas and between plants and gardener remained similar. The key to the gardens lies in the evidence, found throughout books of garden history, that plants were *used* for some particular purpose. 'Abused' might be thought more apt by the ecologically minded. Plants took the place of stone, of brick, of gravel, of ceramic mosaic, of paint or of fabric to produce geometrically intricate patterns. The shears moulded them into weird and fantastic shapes, while the scythe, albeit laboriously and imperfectly, carved grass into lawn. Labour was virtually limitless, which was fortunate, for weeding must have been a nightmare.

In the middle of the sixteenth century, such an environment produced the Tudor garden with its maze and its mount, and the Elizabethan knot. The seventeenth century brought more trees and shrubs into the Jacobean garden, which was followed in turn by the French- and Dutch-influenced stately home gardens, of which many still survive. Here gardening's divorce from nature surely reached its zenith, and if the botanist can find any salvation in such living carpets, it is that Britain was at least spared the most outrageous excesses of the Frenchman, Andre Le Nôtre – Chantilly, Vaux-le-Vicomte, Rambouillet and, of course, Versailles, carved by 36,000 workmen over 16 years. Some ecosystem!

Then, like the first shaft of morning sunlight over the dark hill of formal gardening, came the early eighteenth century and, in Britain at least, two men were warmed by its rays, Lancelot 'Capability' Brown (1716–1783) and Humphry Repton (1752–1818). Their day of work and of immediate influence was fairly brief; its sun set as Victoria came to the throne, but the results of their travail live on in the maturity of our great landscaped parks, and, would that it were fully recognised, in most of our home gardens too. Their role was as shapers of the countryside rather than as gardeners, but they brought a novel and fresh approach to man's relationship with his plants and, in sweeping away, like the leaves of autumn, dozens of formal gardens, went part of the way back towards natural habitats and plant associations. Not, it must be said, all of the way, for the so-called parkland landscape created by Brown and, in modified form, by Repton, has no real natural counterpart; there is no very similar ecological community in a temperate climate of grassland and trees, and many examples of Brown's work have their nearest natural resemblance in some of the tropical savannahs that, of course, he never saw. Nonetheless, to an ecologist, the work of Brown and Repton has two great appeals; first that rigid geometry as a guiding principle of planting was at last abandoned, and second that, at least in part, the armies of gardeners needed to maintain the integrity of Le Nôtre's parterres were replaced by herds of deer and cattle. Lancelot Brown has been called many things from visionary to vandal, although I am unaware of him ever being dubbed ecologist. But, intuitively, ecologist he was; and the grounds of over a hundred British stately homes still bear evidence of this.

Gardening at first ignored many of the lessons and examples of Brown and Repton while the Victorian empire builders set their horticultural fob-watches by the clocks of fifteenth-century Italy. But the time was not so far off when garden plants would also be seen in surroundings more recognisable as ecological associations, and would be grown for their own sakes rather than as living versions of some other artistic medium.

William Robinson (1838–1935), creator of the English garden as so many understand it today, was an Irishman. He was an admirer of much of Brown's work, although well aware that the landscape movement in general had produced some problems. He commenced his seminal gardening publication with a chapter entitled 'Landscape mistakes near the country house' in which he outlined most forcibly the difficulties attendant upon planting together tree and shrub species from differing natural environments. In these few lines, he almost laid the foundations for ecological thought in gardening. He went on to describe, at great length, his abhorrence of garden formality, and his words should be read, in all their virility and with all their vituperation, by every modern gardener. Nonetheless, while Robinson displayed an intuitive ecological mentality, he certainly lacked some of the ground rules needed to put all his notions into practice. He seemed, as in the instance quoted above,

William Robinson 1838–1935
Although an Irishman, Robinson's name is now synonymous with English gardening as so many understand it. Much of the 'naturalism' that he introduced was intuitively ecological.

The Catshead Codlin apple is one of many old forms that owe their survival to cottage gardens.

to grasp the basis of competition but not fully to understand succession. Witness his comment on what he persisted in calling a wild garden, 'the owner might go away for ten years and find it more beautiful than ever on his return'!

It is inappropriate to discuss in detail Robinson's work and influence, for much critical and scholarly analysis has been written on this quite extraordinary man, although no words describe his beliefs better than his own, which I commend to you. I do not ignore that many of his ideas were unoriginal and that he had influential contemporaries. Shirley Hibberd, John Ruskin, William Morris and many others all played their part in overturning Victorian gardening formality; but none seems to me to have felt the pulse of the natural environment as instinctively as did Robinson. And above all, credit must be given to the traditional English cottager's garden in helping to shape Robinson's destiny; in 1883, he wrote '. . . among the things made by man nothing is prettier than an English cottage garden'.

While Robinson thundered with his pen, and poured his ideas with his plants into his own garden at Gravetye Manor in Kent, another force was at work spreading much of his message in gardens across the nation. Robinson may have been an Irishman, but only England could have produced a Gertrude Jekyll. Late in life and with failing eyesight, this remarkable woman met two remarkable men, Robinson himself and the architect Edwin Lutyens. She took Robinson's ideas to heart, discarded the more extreme, practically difficult, aesthetically unsatisfying and, almost intuitively, many of the ecologically unsound, and adapted them above all to form a garden feature that is forever English and forever hers, the herbaceous border. While Lutyens designed the houses and the gardens, Jekyll planted them with vision. The herbaceous border, *à la* Jekyll, really is plant succession translated into gardening terms.

And so to the present century, when few owners have gardens of a size to accommodate even part of a Jekyll border. It is with ingenuity, nonetheless, that her work has been adapted to our very different times, circumstances and demands. First, however, let twentieth-century gardening at all its levels be set into the overall context of twentieth-century life, which influences it and which it, in turn, reflects. The century has been punctuated by periods of international conflict. On the one hand, this has meant that men's minds were preoccupied with more

Attingham Park, Shropshire (now owned by the National Trust); plan and impression by Humphry Repton 1752–1818 from a Red Book of 1797. In the Red Books, prepared for his clients, Repton drew and described his plans and ideas for remodelling their properties. With Lancelot 'Capability' Brown 1716–1783, Repton introduced a quite novel form of landscape into many of our great country estates, although it is a landscape that bears little resemblance to any natural temperate climate plant community.

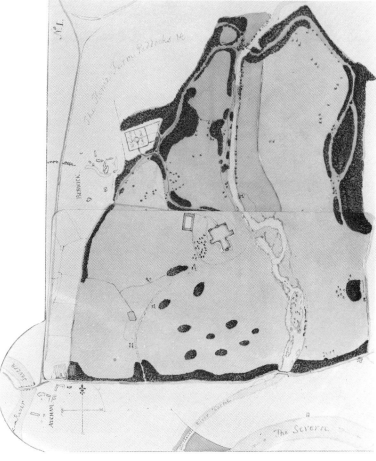

momentous happenings than the planning of herbaceous borders, while on the other the garden itself, in mind and in reality, has given solace. For how many soldiers in the trenches of the Somme were Rupert Brooke's 'thoughts by England given' the thoughts of her summer flowers and her gardens? Less emotively, moreover, the necessity of feeding the population during times of war and economic depression meant returning the home garden to its ancient role as provider of the family's food needs. Since the end of the second world war, the importance of the garden as food source has waxed and waned, the drought years of 1975 and 1976 fairly firmly suppressing the last great period of excitement in vegetable gardening. After several years of increasing home garden vegetable growing (almost a resurgence of the wartime 'Dig for Victory' enthusiasm), so many gardeners lost so many plants and so much work during the watering restrictions of those two hot summers, that they seem never to have been able to summon up the effort again. Now we live in more stable, if still uncertain and troubled times – times when, for most of the British people, hunger, if it exists, is alleviated more by the welfare state than by the allotment. Unemployment may have brought a shortage of money for many, but it is nonetheless true that home (and, of course, garden) ownership has never been higher.

Modern gardens may be more numerous, but they are also much smaller than the ornamental gardens of times past and this has brought about a significant shift in the types of plant now grown, the ways in which they are grown, and the problems attendant upon their cultivation. My main concern, therefore, is with the twentieth-century home garden in its various forms, but it is important to say something of three other types of modern garden too, for it is there above all that the gardener will receive his stimulus and his ideas.

The Botanic Garden has always had a serious scientific purpose in providing a reference collection of the plants of the world, of the plants of a smaller geographical region, or of particular botanical types or groups. Clearly, to accommodate plants of widely differing environmental origins, the botanic garden must take the most assiduous account of their ecological requirements. They have the invaluable attribute today of generally being staffed by the very botanists who have collected the plants from the wild and whose very purpose lies in understanding their biological roles and functions.

Formality in garden style takes many forms, ranging from the sweeping vista that fronts Blickling Hall in Norfolk (above left) to the geometric lawned quad of New College, Oxford (above right) and the clipped yews and straight beds and ornamental walls of Lanhydrock, Cornwall (right).

Botanic Gardens of the old style, best represented in Britain by the one at Oxford University, founded in 1621, tended in the past to be living museums, where, although the plants were of necessity grown in conditions reasonably close to those of their native homes, little account was taken of their arrangement within the habitat. The updating of these gardens to modern standards, with recourse to modern technology and knowledge, has changed this; and although nothing can compare with the impact on a temperate climate gardener of the first time he stands alone in a real tropical forest, I believe firmly that every botanist and gardener, if he is unable to visit the natural homes of his plants, should at least visit a modern botanic garden. There he will see very faithful copies of a wide range of habitats in a very short time. And no amount of verbal instruction or textbooks can convey to a gardener the reason why his house plants fail to flourish, as well as a visit to the Palm House at Kew can!

The Arboretum parallels the Botanic Garden in respect of woody plants (though, indeed, most of the latter include a significant tree and shrub collection). Many of the most comprehensive tree collections in the British Isles, however (The Hillier Arboretum in Hampshire or the more recent John F. Kennedy Arboretum in Wexford, Ireland), tend to be living catalogues, invaluable as such, but teaching relatively little of tree ecology. Even at Kew, Edinburgh and other great botanic gardens, the trees are usually grown as isolated individuals, and this gives no indication of the way they interact naturally with other trees or with vegetation in general. One of the most satisfying of British ecological arboreta is that at Westonbirt in Gloucestershire (its *Acer* collection a glory in October), where it is at least possible to see exotic trees growing in a reasonably well structured woodland.

Ironically, the first British Municipal Garden, planted at Derby in 1840, was called an arboretum; but planted areas in towns, to which the public had access, very rapidly became a mid-Victorian phenomenon and, within thirty years or so, could be found in almost every town and city in Britain. The public park, as such, was very often a small arboretum (or even a small botanic garden) and displayed to the residents plants that otherwise they would rarely see. It very seldom attempted any semblance of natural habitats and was a pale shadow even of the landscaped park of the landed gentry of the time. The typical municipal garden today, however, is a most curious environment, almost a living fossil of mid-nineteenth-century horticulture. It is the last refuge of the formal bedding arrangements otherwise swept away by Robinson, and it consumes annual bedding plants by the million for planting in rigid and often very complex geometric patterns. These gardens bear no resemblance whatever to any natural plant association; their rigid formality is reflected even in the choice of tree with which they are often studded, the South American Monkey Puzzle, *Araucaria araucana*. Nonetheless, the formal municipal bedding scheme clearly gives pleasure to many, and certainly brightens traffic islands and other dull areas during the summer. It shows little signs of giving way to more recent trends, although in the New Towns and in many modern urban estates, designed (or thrown together) by those who rejoice in calling themselves planners, the so-called architectural garden has made its appearance. Here plants, especially trees and shrubs, requiring minimal maintenance and displaying reasonable durability to the vicissitudes of inner city life, are treated, much as building blocks or litter bins are treated, as structural objects. And who is to say that the floral clock, bearing thousands upon thousands of alyssum, lobelia and *Begonia semperflorens* in perpetual motion is any less dignified a fate for a plant than for it to take its place in a circular hole in a slab of pre-stressed concrete as visual relief from a line of lamp-posts?

I use the generic expression 'Country House Garden' for want of something better to describe the modern garden to which the public have access but which is neither botanical nor municipal. It is almost always larger than most home gardens and is very often attached to a large house. The country house garden is where the historical development of British garden style can best be seen, although, thanks to the influences to which I have referred, most such gardens today bear strongly the stamp of Brown, Repton and Jekyll or of their admirers and disciples. Relatively few, even of the fairly recent Jekyll gardens, remain in unaltered state, and present day economics dictate that many have fallen into states of at least partial neglect. The salvation of many has been the National Trust and its regional counterparts, and a visit to a garden under the care of those bodies will at least be rewarded by the evidence of a generally high standard of horticultural attention. So many and so varied are country house gardens, however, that any further generalisations would mislead. I have listed in Table 25, therefore, a range of gardens, with a wide geographical distribution within

The successful planting of rhododendrons (above) depends on the knowledge that they are natural shade lovers and are at home beneath trees, while the herbaceous border (right) demands detailed acquaintance with many different types of plant if it is to succeed in the modern garden.

the British Isles, that are regularly open to the public and that display particularly important features.

Thus we reach the last quarter of the century and the archetypal, present day British home garden. It takes several forms, sometimes graced with the names cottage, suburban, architectural, or even, more recently, wild. It is typically a garden of under 200 square metres, a garden whose overall form, content and management bear to a greater or lesser degree the influences of many centuries of garden use and design, of plant collectors and plant introducers, of scientists, chemists, engineers and technologists and which mirrors a great deal of the overall life and life style of its age. Into this age have come many trends, of which one of the most conspicuous is that I alluded to in Chapter 1; what has popularly become known as the 'Ecological Movement'. I define this as a general concern for the fate of the natural environment at human hands; only a small part, of course, of ecology as a whole. It has become confusingly and fairly inexplicably associated in many minds with nuclear disarmament and predominantly left wing politics. If a finger can be placed on a seminal date for the beginnings of this movement, it is probably 1962 and the publication of 'Silent Spring',

Knightshayes, Devon, a fine example of a modern garden in which trees grow in at least a semblance of natural woodland.

Rachel Carson's wide ranging book, over-generalised and occasionally inaccurate, but entirely justifiably drawing to the world's attention the then trend towards ever more use of artificial, and especially persistent, pesticides. The ripples from Rachel Carson's splash are still to be felt and the restrictions placed in 1985 on the use of the herbicide ioxynil in British garden have, indeed, a long pedigree.

Something of an orphan that has been embraced, not to say swallowed whole, by the more widespread concern for the environment, has been what is usually called organic gardening. I have discussed elsewhere in the book what I believe to be some of the misguided tenets of this fraternity, especially with regard to their views on what constitute acceptable and unacceptable types of fertiliser; but with their overall aims and undoubted sincerity, I can readily identify. In Sir Albert Howard (1873–1947), organic gardening had its guru (and I use the word advisedly, for many of his ideas were established among the Indian peasantry), but the theme now has its own inertia and has, most interestingly, at last made its presence felt in commercial horticulture and farming.

Thus we have a climate of increasing public awareness of the possible hazards resulting from abuse of artificial chemicals, and a trend among

The municipal garden and its emulators represent what some have called fossil horticulture; almost the last survivors of the old carpet bedding approach to the growing of flowers.

much of the gardening movement towards at least partial reliance on composts, manures and 'organic' products derived more or less directly from some once living organism.

Compound this state of affairs in the typically small home garden with an inherent native addition to collecting and plant variety (as Anthony Huxley has said, 'most British gardeners are too much interested in plants to accept a reduction in numbers'), a modern marketing environment that has produced the garden centre and pre-packed garden requisites (including plants), increased leisure time (although not necessarily the wish to spend it all on gardening), and a high level of dependence on the supermarket for food, and you have some idea of the modern British gardening environment.

What has been the significance of these factors on the ecology of the modern home garden and on the ease with which it can be managed? Given as his customer the owner of a small garden, with a wish for a great many plants and a minimal labour input, the plant breeder has produced one of the real phenomena of post-war gardening, the 'compact variety'. Shorter-stemmed herbaceous perennials, requiring little staking are now *de rigueur* for garden centres. Sadly, the dwarf cultivar must still compete with the full-sized weed and has contributed therefore to the popularity of the second of the modern horticultural desiderata, the ground-cover plant. Low-growing, if not prostrate, claimed (at least by the people who sell them) to suppress weed growth, these are plants that in seeming to exemplify the best of ecological gardening have captured the public's imagination. Unfortunately, they too often fall short of expectations: there are very, very few plants indeed that fulfil all of the hoped-for criteria. The weeds will still grow while the infant slow-growing plant takes its many years to reach maturity, whereas the faster establishing and maturing species are quite likely to subjugate, along with the weeds, the other garden plants whose well-being they are intended to aid. Ultimately, many plants so widely recommended as ground cover (and the infamous *Hypericum calycinum* must top this list) will, in most modern gardens, become a greater menace than any annual weeds and, even then, present little threat to many of the most troublesome perennials.

In a great garden, with great space, there is no difficulty in making a choice of plants to give continuity of flowering through twelve months. The small home garden presents immeasurably greater problems, how-

Mutilated street trees are testimony to the poor choice of species so often made for particular habitats. It is better to remove a tree that proves too large than subject it to hideous abbreviation.

ever, and the third of the modern gardening trends has been to pay greater attention to the merits of leaves for garden ornament. I have long held the belief (by no means original) that the two most important colours in a garden are green and white, but to have expected a general nursery to offer over thirty different types of *Hosta* (and even for there to be a society specialising in their cultivation) would have been un-thinkable until recently. The immense variety in leaf form among temperate climate plants offers greater scope still for the future, provided the gardener heeds at least some of the ecological reasons for the evolution of these forms and provides appropriate garden conditions for them.

The ecology of Manhattan Island has taught the modern gardener a lesson too: with limited space, you must go upwards, hence the increasing attention paid by gardeners and their suppliers to climbing plants. Words of caution must be voiced in this regard, nonetheless, for the natural habitat of the climbing plant is not restricted by the top of a house wall or a fence; many are extremely vigorous. Many are plants of shaded woodland and may have the merit of flowering early or late in the season, so avoiding the period of maximum canopy cover, but conversely, may look dismal for the remainder of the year.

Limited space in the modern home garden has provided the impetus for ever more use of pots, tubs and other containers, for these can be placed on path edges, terraces and patios to increase the available growing area appreciably. But they do not constitute the ideal system for the gardener with limited time, because they need additional watering to compensate for being denied the opportunity to tap water reserves from a natural soil.

The modern home garden, therefore, utilises many of the basic principles of ecology in seeking to accommodate the very special demands of its owner; but the further one moves from the natural plant habitat (and there are undoubedly some modern gardens that seem to have come almost as far as those of Le Nôtre), the greater are the difficulties in managing it properly. Ultimately, I must conclude that if the small modern garden really is a labour-saving garden, it is more because it *is* small and because a gardener can work it thoroughly in a few hours, than because of the ecological theory of the ground cover, the compact variety or any of the other noble principles.

So if today, we are a nation of gardeners, are we also a nation of ecologists? Are the two disciplines siblings or merely cousins? I have

The National Trust and its regional counterparts have been the salvation of many fine old gardens like Hidcote in Gloucestershire.

232

Among the most popular types of plant today are the weed-suppressing ground cover varieties such as (above top) the winter heather 'Springwood Pink'. Also popular for the small modern garden is the dwarf variety – miniature roses like 'Magic Carousel' seem to meet many of the hoped-for criteria of long flowering season, ease of cultivation and low growing habit. The vegetable plot (above right) is still the core of many people's gardens although vegetable growing in general suffered a setback in the drought years of 1975 and 1976 from which it has never really recovered. Some of the prettiest gardens and best use of limited space can be found in towns and cities. This town garden (right) represents a haven of tranquility among the noise of streets and traffic.

pinpointed the origins of modern British gardening in the late nineteenth century, with the work of Gertrude Jekyll and William Robinson and, if pushed, I would actually stick the pin in the calendar for the year 1883 and the publication of Robinson's 'English Flower Garden'. To pinpoint the beginnings of ecological science is also a subjective matter, but I would opt for 1896 and the appearance of Warming's 'Lehrbuch der Oekologischen', for although vegetational succession, for the comprehension of which I owe most to Warming, is far from ecology *in toto*, it is the germ-centre for what I understand as ecological thinking. I doubt whether Robinson read Warming or vice versa and, indeed, if Robinson had thought entirely ecologically, he would not have made some of the more outrageous of his statements (although his gardening heart was clearly in the right place). Over many years of observing gardeners and ecological botanists, I have come to the conclusion that the twain rarely meet (hence this book's *raison d'être*) and that, where gardeners like Robinson do seem to have been taking account of ecological principles, it has been a classic instance of convergent evolution.

Unless he has made an accidental or deliberate change of career direction, the modern professional gardener's training is invariably in horticulture, not botany. And yet, few horticulture courses that I know of include a significant, or indeed any, component of ecology. In general, the gardener is not trained or encouraged to think ecologically and yet, as I have stressed several times, the truly skilled gardener or garden designer is an intuitive ecologist. If he did not take account of such sound ecological principles as competition, succession and invasion, his gardens would fail aesthetically and functionally. The ecologist, I have often thought, is the true four-dimensional biologist, and thus the skilled garden designer must be the four-dimensional artist. Anybody can draw a one dimensional dot; most people can produce some semblance of a two dimensional drawing; many can, with tolerable accomplishment, concoct a three dimensional sculpture. Yet to paint, like Gertrude Jekyll, in four dimensions, and in colour, is a rare art indeed.

Gardening, then, most assuredly is plant ecology under new management, but while ecological theory can explain each and every gardening facet and operation, it requires vision, patience, experience, understanding and thought to put into practice the innumerable possibilities that it offers. I hope that among these pages I have provided the thinking gardener with a little more food for just such thought.

Table 25 A few British gardens open to the public that display significant ecological and historical features

Garden	Approximate period of major development	Principal designer(s)	Major features
Arley Hall, Cheshire	1830–60	Rowland Egerton-Warburton	Some of the largest, oldest and finest herbaceous borders in England, well known to Gertrude Jekyll
Blickling Hall, Norfolk	1600–1900	Nesfield & Digby Wyatt	Elaborate Victorian parterres with lawns, geometric beds, fountains and clipped yew
		Lyminge & William Ivory	Woodland Garden enclosed by raised terraces with avenues of trees and rhododendrons
		Probably Humphry Repton	Landscaped park with lake
Drummond Castle, Perthshire	Early 19th century	Uncertain	Spectacular terraces and parterres with strong Italian influences
Hatfield House, Hertfordshire	1500–1980	Thomas Chaundler and others	Restored Elizabethan and Jacobean gardens with 400 years of other influences
Hidcote Manor, Gloucestershire	1907–58	Lawrence Johnston	'Garden rooms' – several individual gardens in different styles, bounded by hedges
Knightshayes Court, Devon	1870–1960	Amory family	The archetypal English country house garden, blending the formal and informal
Levens Hall, Cumbria	1700	Guillaume Beaumont	Probably the most spectacular yew topiary in England
Plas Newydd, Gwynedd	1800 and later	Humphrey Repton and others	Landscaped park surrounding classic country house gardens with Italian and other influences
Royal Botanic Gardens, Kew, Surrey	1720–present	Many	As a 'garden', a hotch-potch but with the finest collections of plants in the British Isles and arguably the most beautiful glasshouses in the world
Westbury Court, Gloucestershire	1700	Maynard Colchester	Classic 17th century Dutch water garden; a very rare survival of the period
Westonbirt Aboretum, Gloucestershire	1830–1930	W. S. Gilpin and others	The finest collection of trees in Britain in what passes for an approximately natural association

Palm House
Royal Botanical Gardens, Kew

Index

Publisher's Acknowledgements

All of the diagrams in this book were produced from data supplied by the author. In most cases this information came from several sources, and it is impossible to identify them all individually. In a few instances, where a diagram was based fairly closely on another, this is acknowledged in the caption.

The Soil Map on p. 51 is reproduced from the 1:63,630 Soil Survey Map of England and Wales by kind permission of the Head of the Soil Survey of England and Wales. The Geological Map on the same page is reproduced from the 1:63,630 Geological Survey Map of Britain with the permission of the Controller of Her Majesty's Stationery Office, Crown copyright reserved. The Land Use Map on p. 119 is reproduced from the 1:25,000 Land Use Sheet for Coldstream, with the permission of the Controller of Her Majesty's Stationery Office, Crown copyright reserved; land use data is copyright reserved by the Second Land Utilisation Survey of Britain. The Francis Dodd etching of William Robinson on p. 220 is reproduced by kind permission of Miss Grace Woodbridge, and supplied by Mr Peter Herbert. It first appeared in the book by the late Mea Allan *William Robinson, Father of the English Flower Garden* (Faber and Faber Ltd, 1982). The plan and map of Attingham Park by Humphry Repton on page 223 are reproduced with permission from the Royal Horticultural Society Lindley Library.